Pondweeds, Bur-reeds and their Relatives of British Columbia

Aquatic Families of Monocotyledons

Pondweeds, Bur-reeds and their Relatives of British Columbia

Aquatic Families of Monocotyledons

T. Christopher Brayshaw

Published by the Royal British Columbia Museum, 675 Belleville Street, Victoria, British Columbia, V8W 9W2, Canada.

Illustrations and maps by T.C. Brayshaw.

Edited by Gerry Truscott (RBCM) and Susan Clarke.
Designed and typeset by Gerry Truscott in Times 10/12.
Cover design by Chris Tyrrell (RBCM).

Printed in Canada by Friesens.

Canadian Cataloguing in Publication Data
Brayshaw, T. Christopher, 1919-
Pondweeds, bur-reeds & their relatives of British Columbia

Previously published by British Columbia Provincial Museum, 1985.
Includes bibliographical references: p.
ISBN 0-7718-9574-7

1. Aquatic plants - British Columbia. 2. Monocotyledons - British Columbia.
I. Royal British Columbia Museum. II. Title.

QK203.B7B722 2000 584'.176'09711 C00-960342-5

CONTENTS

PREFACE TO THE SECOND EDITION

A number of changing situations has made a revision of this book necessary: the discovery of species and varieties not previously known to occur in British Columbia; revisions to the identities and names of several plants; and adjustments to the system of classification.

Adjustment of classification systems is a continuing process, as the development of new and progressively more precise techniques, such as chemical and molecular analyses, lead to refinements of our understanding of the relationships among families and species. A classification scheme, as published in a book, resembles a single frame in a moving-picture film. It will be followed by other frames as the movie progresses.

In addition to revising the text, I updated the range maps, revised many of the figures and drew several new figures.

I wish to thank John Pinder-Moss, collections manager in the Royal British Columbia Museum herbarium, and Dr Adolf Ceska for their advice and help in finding plant records and references, and the curators and staff of the University of Victoria and the University of British Columbia for their help in finding material for my examination. Special thanks to Dr Richard Hebda, Curator of Botany at the Royal B.C. Museum, for his encouragement and for reviewing this manuscript.

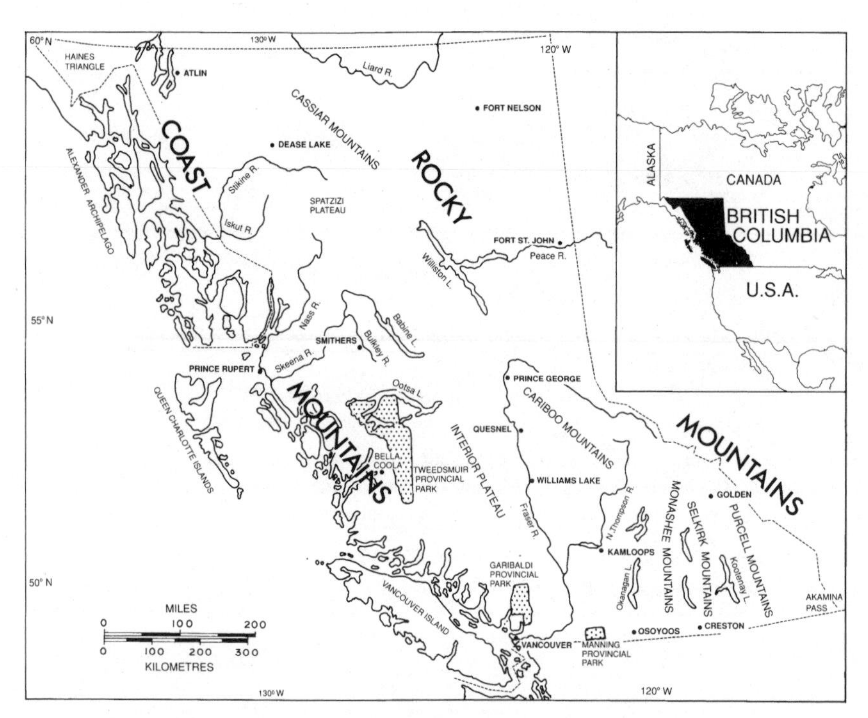

General geographic features of British Columbia.

INTRODUCTION

Definition of the Group

This work is primarily taxonomic and descriptive in scope. The plants described fall into one class of flowering plants, the monocotyledons, and comprise those families all of whose British Columbia representatives are aquatic or amphibious.

All the members of the families treated here are herbaceous, and show various kinds and degrees of modification for life in aquatic environments. In other respects, while often sharing few conspicuous physical features in common, these species show their relationships in anatomical details often of microscopic scale.

Many systems of plant classification have been assembled over the years. But, up to now, no system proposed enjoys universal acceptance as a natural scheme. In this book, I use a system that combines elements of those in Dahlgren, Clifford and Yeo 1985 and Cronquist 1988. Other, more natural systems, better reflecting the true genetic relationships among plants, will undoubtedly appear in the future, as active research currently underway improves our knowledge of these interesting organisms.

Among the dicotyledons (the other class of flowering plants), aquatic species and genera are scattered through a large number of families, few of which are entirely aquatic. On the other hand, most of the plants described in this work fall into nine families in one superorder, the Alismatiflorae, all of whose members (including three families not represented here) are amphibious or aquatic. Of the remaining five families treated here, four are entirely aquatic or amphibious, while the Araceae (the Arum Family), whose three British Columbian species are aquatic, are mostly terrestrial in other parts of the world.

As treated here, our 14 families are grouped in five orders, in four superorders in the system of Dahlgren, Clifford and Yeo 1985 and in three subclasses of the system of Cronquist 1988.

The hierarchical checklist following this section (on page 5) indicates how the species and their varieties described here fall into their families and larger categories. The positions and relationships of these families and larger categories in the overall classification of the monocotyledons are indicated in figures 1A and 1B.

Four families of the monocotyledons that contain some aquatic members are omitted here: the Gramineae (or Poaceae), the grasses, of the Order Graminales (or Poales); the Cyperaceae (the Sedge Family) of the Order Cyperales, the Juncaceae (rushes) of the Order Juncales; and the Iridaceae (irises) of the Order Liliales. The grasses and sedges are subjects of Royal British Columbia Museum handbooks (both out of print, but available in many libraries); the rushes need further work, and probably should be the subject of a separate book; and the irises are mostly terrestrial, with two amphibious species in British Columbia: *Sisyrinchium californicum* (Golden-eyed-grass) and the introduced *Iris pseudacorus* (Yellow Flag).

Collections

Aquatic plant species are seldom as well represented in collections as terrestrial plants. Among the reasons for this situation are the unobtrusiveness of many of the submerged plants, which remain out of sight to observers standing on dry land, the necessity of having a boat for adequate sampling of open-water plant communities, and the awkwardness and discomfort of sampling swampy or marshy vegetation types. As a result, people have made few collections of our aquatic species, and our knowledge of our aquatic flora is correspondingly weak.

In an attempt to fill the gaps in our knowledge of these plants, I took a boat or canoe on several field trips and concentrated on sampling aquatic sites distributed widely across British Columbia. The resulting collections are filed in the herbarium of the Royal British Columbia Museum.

I also examined aquatic plant collections in herbaria at the University of Victoria, the University of British Columbia (Vancouver), the National Museum of Natural Sciences (Ottawa) and the Biosystematics Research Institute of Agriculture Canada (Ottawa), as well as a large collection of aquatic plants made by the staff of the British Columbia Ministry of Environment, Lands and Parks (Victoria). I used the records in these collections to draw the maps of species distributions in British Columbia.

For the brief statements of species distributions beyond the borders of British Columbia, I consulted several regional floras, including Calder and Taylor 1968, Cody 1996, Dandy 1980, Gleason and Cronquist 1991, Hitchcock 1969, Hulten 1941 and 1968, Porsild and Cody 1980, and Scoggan 1978.

CHECKLIST

The following is a hierarchical checklist of the families, genera and species dealt with in this book.

Subclass Alismatidae
  Superorder Alismatiflorae
    Order Alismatales
      Family 1: Butomaceae
      Butomus umbellatus L.

      Family 2: Hydrocharitaceae
      Egeria densa Planchon
      Elodea callitrichoides (Richard) Caspary
      E. canadensis Richard
      E. nuttallii (Planchon) St John
      Vallisneria americana Michaux
      V. spiralis L.

      Family 3: Alismataceae
      Alisma gramineum Lejeune
        var. *angustissimum* (DC) Hendricks
        var. *gramineum*
      A. lanceolatum Withering
      A. plantago-aquatica L.
        var. *americanum* Schultes and Schultes
        var. *plantago-aquatica*
      Sagittaria cuneata Sheldon
      S. latifolia Willdenow

    Order Najadales
      Family 4: Scheuchzeriaceae
      Scheuchzeria palustris L.
        var. *americana* Fernald

Family 5: Juncaginaceae

Lilaea scilloides (Poiret) Haumann
Triglochin concinna Davy
  var. *concinna*
  var. *debilis* (M.E. Jones) J.T.Howell
T. maritima L.
T. palustris L.

Family 6: Potamogetonaceae

Potamogeton alpinus Balbis
  var. *subellipticus* (Fernald) Ogden
  var. *tenuifolius* (Rafinesque) Ogden
P. amplifolius Tuckermann
P. crispus L.
P. diversifolius Rafinesque
P. epihydrus Rafinesque
  var. *epihydrus*
  var. *ramosus* (Peck) House
P. filiformis Persoon
  var. *alpinus* Blytt
  var. *filiformis*
  var. *occidentalis* (Robbins) Morong
P. foliosus Rafinesque
  var. *foliosus*
  var. *macellus* Fernald
P. friesii Ruprecht
P. gramineus L.
  var. *gramineus*
  var. *maximus* Morong
  var. *myriophyllus* Robbins
P. illinoensis Morong
P. natans L.
P. nodosus Poiret
P. oakesianus Robbins
P. obtusifolius Mertens and Koch
P. pectinatus L.
P. perfoliatus L.
  ssp. *perfoliatus*
  ssp. *richardsonii* (Bennett) Hulten
P. praelongus Wulfen
P. pusillus L.
  ssp. *pusillus*
  ssp. *tenuissimus* (Mertens and Koch) Haynes and Hellquist

P. robbinsii Oakes
P. strictifolius Bennett
P. vaginatus Turczaninow
P. zosteriformis Fernald
Ruppia maritima L.
var. *longipes* Hagstrom
var. *maritima*
var. *occidentalis* (Watson) Graebner
var. *spiralis* (L. *ex* Dumortier) Moris

Family 7: Zosteraceae
Phyllospadix scouleri Hooker
P. serrulatus Ruprecht
P. torreyi Watson
Zostera japonica Ascherson and Graebner
Z. marina L.
var. *latifolia* Morong
var. *marina*
var. *stenophylla* (Rafinesque) Ascherson and Graebner

Family 8: Zannichelliaceae
Zannichellia palustris L.

Family 9: Najadaceae
Najas fexilis (Willdenow) Rostkovius and Schmidt

Superorder Ariflorae
Order Arales
Family 10: Araceae
Acorus americanus (Rafinesque) Rafinesque
Calla palustris L.
forma *palustris*
forma *polyspathacea* Victorin and Rousseau
Lysichiton americanum Hulten and St John

Family 11: Lemnaceae
Lemna minor L.
L. trisulca L.
Spirodela polyrhiza (L.) Schleiden
Wolffia borealis (Englemann) Landolt and Wildi
W. columbiana Karsten

Subclass Zingiberidae
Superorder Bromeliiflorae
Order Pontederiales
Family 12: Pontederiaceae
Pontederia cordata L.
Zosterella dubia (Jacquin) Small

Subclass Commelinidae
Superorder Typhiflorae
Order Typhales
Family 13: Typhaceae
Typha angustifolia L.
T. x *glauca* Godron
T. latifolia L.

Family 14: Sparganiaceae
Sparganium americanum Nuttall
S. angustifolium Michaux
ssp. *angustifolium*
ssp. *emersum* (Rehmann) Brayshaw
var. *chlorocarpum* (Rydberg) Brayshaw
var. *emersum*
var. *multipedunculatum* (Morong) Brayshaw
S. eurycarpum Engelmann
var. *eurycarpum*
var. *greenei* (Morong) Graebner
S. fluctuans (Morong) Robinson
S. glomeratum Laestadius
S. hyperboreum (Hartman) Fries
S. natans L.

THE AQUATIC ENVIRONMENT

General Types of Aquatic Vegetation

There is a rich diversity of aquatic habitats in British Columbia, and their classification is beset with unresolved complexities. Even circumscribing the application of the term "aquatic" can be the subject of debate, and is, in the end, decided arbitrarily. Generally, the term may be applied to habitats having a permanent water surface, or subject to seasonal flooding or saturation.

Quite apart from the broad regional diversity, with its corresponding diversity of macroclimates, there are varied local wet habitats conditioned by topographic situation, freshwater or marine environment, type and speed of water movement, salinity, acidity, nutrient availability, etc.

There are continuous gradients in the determining factors by which aquatic habitats intergrade with each other and with terrestrial habitats. Thus, a certain degree of arbitrariness must be assumed in drawing up any scheme for their classification. I have simplified, from Zoltai 1976, the following general habitat classification. All these classes of wetlands can be further subdivided according to local situations or conditions.

Open Water

Any habitat covered by water for more than 75% of its area is open water. Zoltai sets a depth limit of two metres, but since many aquatic plants have their bases more than two metres below the water surface, this seems an unnecessary restriction.

Bog

A bog is a spongy, mossy habitat usually dominated by species of *Sphagnum* moss. Typically, the water is held in the moss and in its undecayed remains, is acidic, and is low in available nutrients. The accessible soil is peat, derived principally from the accumulation of dead moss remains and roots. The underlying mineral soil is inaccessible to the roots of plants. Crowberry, ericaceous shrubs (Heath Family), sedges and insectivorous plants are characteristically present; and trees, such as Black Spruce and Tamarack, may be present as open stands of small stature.

Swamp

A swamp is a wooded wetland that is not highly acidic or dominated by *Sphagnum* moss. Commonly bordering lakes or streams, it may be dominated by trees or tall shrubs, such as willows, alders or shrubby birches.

Fen

A fen is an open wetland with a consolidated surface, dominated by sedges or grasses, often with linear patterns of sparse shrubs or trees, drainage channels and ponds.

Marsh

A marsh is an open, treeless site with an unconsolidated surface, and a herbaceous cover of grasses, sedges, bulrushes, cat-tails or other plants, often in clumps among drainage channels and ponds. This class includes the tidal marshes of the sea coast, as well as freshwater marshes.

Community Succession

In water, as on land, nothing is static. All natural communities undergo changes in the course of years, centuries or millennia. As environmental conditions at a site change, through geologic or prolonged climatic changes, or through changes induced by the plants themselves, the species initially occupying the site are replaced by others that are better adjusted to the altered conditions. The process continues until the first community has been completely succeeded by another. This process is known as *succession*.

The succession that follows the usually slow and irreversible geological processes of weathering, erosion and deposition is called *primary succession*. Superimposed on the primary successional trends are short-term, often cyclic, modifications caused by incidental, temporary disturbances such as flooding, fires, land clearing or other human activities. Such changes, which are normally reversible, and often rapid, are called *secondary succession.*

A community reacts upon itself by moderating the basic physical conditions, such as the dominant plants providing shade and shelter and depositing litter. This affects the selection of succeeding species and is often an important factor in determining successional trends, especially in secondary succession.

Aquatic Community Succession

It is widely understood that a lake is a trap for sediment carried into it from surrounding land. Deposition of inorganic silt may be supplemented by that of organic debris where, in an oxygen-deficient environment, organic material can accumulate faster than it decomposes. In this way, a lake becomes shallower until eventually the solid bottom lies exposed to the air. At successive stages in this physical process, successive communities of plants, each adapted to a certain range of water depth, occupy a site, until open water communities give way to marsh, fen or swamp communities.

Succession in the opposite direction may also occur in cold regions. Where soil drainage becomes impeded, especially in areas of low relief, the water table may rise to the surface and *Sphagnum* mosses may assume dominance of the saturated site. The slow oxidation of their remains under these conditions leads to the accumulation of peat, and the eventual replacement of forest cover by a boggy landscape of peatlands and ponds.

Lakes and streams in settled districts are inevitably affected by human activities, and secondary successional changes are extensively altering the primary succession of natural communities. A type of secondary succession that is becoming ever more remarked on is that caused by pollution of waters by material of terrestrial origin. The addition of nutrients in solution or suspension, including nitrogenous substances and phosphates from sewage or from fertilizer-laden runoff from cultivated land, alters the chemical nature of the aquatic environment. This provides special favours for those species that can take advantage of the nutrients and indulge in excessively vigorous growth at the expense of those species that are unable to benefit from the enriched environment. By the shading or crowding out of the unaided by the adaptable species, a new community is selected out of the former one; and the slow succession of the original aquatic communities is redirected, and accelerates in new and often unpredictable directions.

Interactions Among Aquatic Plants

The interactions among members of aquatic communities are often very complex. Competition for light, oxygen, and dissolved nutrients plays an important part. Competition for light, with its secondary effects on oxygen availability and temperature gradients, is the most easily visualized.

Although deeply submersed plants may obtain sufficient light for photosynthesis in clear water, the growth of densely leafy plants with foliage at or near the water surface, often accompanied by growths of suspended or attached algal colonies, can drastically reduce the amount of light reaching plants at greater depths. The effect is analogous to that in a deep forest, where the lower strata of a multilayered community must tolerate the deep shade cast by tall, canopy-forming trees.

Plants with broad floating or emergent leaves are usually confined to sheltered bays and ponds, since they are mechanically ill-adapted to withstand the impact of heavy wave action. They may so shade the underlying water that very little in the way of submersed plant life can grow beneath them. But submersed species may be well represented in waters that are too deep or too exposed to be dominated by floating-leaf forms.

As a secondary effect, the reduced light intensity in deep water results in a reduced rate of photosynthesis and oxygen production by deeply submersed foliage. This can lead to oxygen-deficiency in deep water, since the lower parts of the plants produce no more oxygen than they demand at night and direct solution of atmospheric oxygen in the water is inhibited where the surface is covered by floating leaves.

An oxygen shortage at depth can affect aquatic life not only directly, but indirectly, because the decay of organic detritus, though largely anaerobic, induces an oxygen demand in excess of the available supply. Such a situation becomes habitable only for anaerobic bacteria, with the consequent production of methane, hydrogen sulphide and, possibly, other more toxic end-products.

By impeding the penetration of sunlight to deep waters and obstructing water circulation, a wealth of foliage at or near the surface leads to exaggerated temperature differences with depth, and enhances the stratification of the water. Vertical mixing does not occur – the surface water becomes warm and the deeper remains cool (Dale and Gillespie 1977).

A consequence of the relatively heavy reliance by aquatic plants on vegetative reproduction compared to sexual reproduction (see page 18) is the frequent occurrence of extensive clones – dense colonies of genetically identical "individuals". This kind of colony, once firmly established, may dominate an area to the almost complete exclusion of other potential competitors, and continue to dominate long past the point of that species" optimum adjustment. Under these conditions, invading members of a later successional stage must not only be adapted to the changing physical environment, but must be able to compete with the old established clones of earlier successional species. This situation undoubtedly has a restraining influence on the rate of community succession.

Though interactions with animals are not within the intended scope of this work, it may be pointed out that many of the plant species dealt with here are important food sources for waterfowl and aquatic rodents such as muskrats (Fassett 1957). Waterfowl, in their turn, are important dispersing agents for many of these species.

The Aquatic Environment as a Habitat for Plants

Every land plant must deal with the management of its water resources. It has to obtain water from an often irregular source of supply and then conserve against excessive loss by evaporation. The need to facilitate the exchange of oxygen and carbon dioxide between the internal tissues and the atmosphere imposes a need for mechanisms that compromise the retention of water against evaporative loss.

An aquatic plant, with its roots, and often all of it, immersed in water, is spared the problem of maintaining its most urgently needed resource. But relief from this stress is obtained at a price. Aquatic plants must deal with stresses and hazards that land plants are not exposed to, and they must evolve special adaptations to successfully invade and exploit a watery environment.

The environmental factors presented by a fully aquatic habitat are:

Mechanical

Water provides support through buoyancy, making sturdy, rigid stems and woody tissue unnecessary. Rigid stems might buckle or break under the stress of river currents, tidal currents or wave action, making stiff or woody tissues a disadvantage. Heavy wave action causes abrasion of foliage and stems against rocks, and necessitates continual replacement of the lost tissues.

Gas Exchange

Oxygen and carbon dioxide are just moderately soluble in water, and diffuse much more slowly in water than in air. This becomes a serious limitation for supplying oxygen to, and removing carbon dioxide from, rhizomes and roots that are buried in unaerated mud at the bottom of a water body, or merely deeply submerged where the water is stagnant. Even in marshes or fens, water-saturated, unaerated soil inhibits the diffusion and supply of oxygen to buried roots.

Opacity of Water

Though intrinsically transparent, water often carries material in solution or suspension that absorbs incoming light, so that insufficient light reaches the deeper water to support plant growth. Waters of this kind, as in many of our silt-laden streams, are usually sterile of vegetation, except where very shallow.

Fluctuating Water Levels

The unreliability of the main resource – water – is emphasized in the alternate flood and drought that affects water bodies, such as reservoirs with deep seasonal draw-downs. In rivers such as the Fraser, the water is high and turbid in summer, and low and clear in winter. Any plants submerged during the summer growing season at the limit of light penetration would be exposed to drought and frost in winter. Such an environment is inhospitable to flowering plants and is inhabited, if at all, by a few lichens or mosses.

Interference with the Traditional Pollination System

Flowering plants are pollinated traditionally by insects; and wind pollination is common in situations where insects are less effective. Aquatic environments interfere with pollinating systems dependent on insects; the more aquatic the habitat, the more extreme the interference. Flooding or frequently repeated inundation by waves or tidal action can render the normal pollinating vectors of terrestrial flowering plants – insects or even wind – inefficient or totally inoperative. An aquatic environment requires plants to adapt pollinating systems for this special situation.

Seed Dispersal and Reproduction

The dispersal of fruits or seeds from a water surface by wind would be risky and the distance conveyed short. Thus, we see few wind-borne fruits or seeds in aquatic plants; the notable exception in this group being *Typha*, which carries its fruiting spikes usually a metre or more above the water. Dispersal by land birds is likewise hindered, since they want perching places near their food source, something commonly lacking in extreme aquatic situations. Few aquatic plants produce berries that are attractive to land birds. An exception is *Calla*, which normally grows attached to the shore, or among emergent shrubs, and holds its fruiting spike above the water. The foraging habits of aquatic birds and mammals have provided different opportunities for growth and reproduction in aquatic plants.

EVOLUTION IN AQUATIC PLANTS

Adaptations of Plants for Aquatic Life

The aquatic plants treated herein display a fascinating diversity of modifications that adapt them to their environment. Starting from an original more-or-less terrestrial precursor with erect form and complete, showy flowers pollinated by insects (the basic pattern in flowering plants), progressive modifications may involve any one or more of the following changes:

Reduction of Supporting Tissue

In a fully aquatic environment, significant material and synthetic effort can be conserved through the reduction or loss of xylem (woody vascular tissue). The xylem's rigid supporting function is replaced by the buoyancy of the water, and apart from the secondary function of transporting absorbed nutrients about the plant, its water-conducting function is superfluous to a plant that is immersed in water. Stiffening by lignification of such xylem cells as remain may be reduced or even eliminated.

Attainment of Flexible Strength

The development of long pliable stems or leaves enables plants to bend to water currents without breaking. This is largely achieved through the concentration of the vascular tissues of a stem into a slender, central, poorly lignified strand, having tensile strength but no rigidity. Where extra tensile strength is then needed, as in the surf-grasses (*Phyllospadix*), the vascular strands are supplemented by numerous very fine strands of sclerenchyma, whose much-thickened cellulose cell walls provide strength without a conducting function.

Replacement of Abraded Tissues

Fast water currents and wave action sweep branches and leaves against rock or other hard surfaces, causing abrasion of the tips of leaves and stems. An important aspect of growth here lies in the activity of basal meristems. These are special growing tissues, widely found in the monocotyledons, at the bases of internodes of stems and at the bases of leaf blades. These meristems produce new stem or leaf tissues, thus renewing these organs from their relatively protected bases as the tips are eroded away. In aquatic monocotyledons, these meristems are commonly enclosed and protected by the basal sheaths of adjacent older leaves.

Development and Function of Lacunate Tissue

Lacunate tissue, spongy tissue containing interconnecting air passages, may serve two purposes: buoyancy, to keep foliage near or at the surface of the water; and improved gas exchange in deep-lying tissues, allowing oxygen to reach buried rhizomes and roots, and carbon dioxide to leave them, since these gases diffuse much faster through air than through water.

In at least some aquatic plants, the circulation of gases by diffusion is augmented by mass flow driven by modestly elevated pressure. In some plants with floating or emergent leaves, active young leaves have been found to absorb air from the atmosphere, and force it down the lacunae in the petioles, using the energy of the sun's heat (Dacey 1980). In this way, oxygen in excess of that of photosynthetic origin is supplied to the buried plant parts. A return flow, enriched with carbon dioxide, makes its way to the surface via older leaves. Even in submerged foliage, oxygen from actively photosynthesizing young foliage may flow through the stem lacunae under slightly elevated pressure.

Lacunate tissue is commonly formed by extreme dilation of the intercellular air spaces that normally occur in the cortex and leaf mesophyll that lie outside the vascular strands; but air passages may also be formed by partial dissolution and replacement of the xylem that would be present in the vascular strands of terrestrial plants.

Tolerance of Oxygen Shortage

The tissues of roots and rhizomes buried in deep, oxygen-deficient environments, such as the mud or muck of a lake bottom, possess physiological adaptations that enable them to survive periods of oxygen deficiency. Under such conditions, these tissues can replace the normal oxygen-dependent respiration process with an oxygen-independent fermentation process, forming methanol or lactic acid, rather than carbon dioxide, as a waste product. This process is important during winter and at night, when darkness cuts off the supply of oxygen from photosynthesis.

Reduction of Leaf Tissues

Reduction of leaf tissue is common, especially in submerged leaves. Membranous and very thin, threadlike, or finely dissected (in dicotyledons), leaves provide a large surface area for gas exchange in relation to the volume of the leaf, which counteracts the effect of the low solubility and slow diffusion of gases in water. In this way the internal photosynthesizing tissues (mesophyll) are in closer communication with the surrounding water. The complex mechanisms for controlling water loss in aerial leaves are unnecessary under water. Stomata may be non-functional or absent. The epidermis may contain chlorophyll – in contrast to its usual character in land plants – and it may replace most or all of the photosynthesizing function of the reduced or absent mesophyll. Thus the photosynthesizing chlorophyll is brought as close as possible to the external source of carbon dioxide.

Reduction of Roots

Adaptation of aquatic life in still water may involve the reduction, even to complete elimination, of the root system, since its primary function here is for anchorage, not water absorption. Root hairs are often dispensed with. The extreme condition of this adaptation leads to the detachment of the plant from the solid bottom substratum to become freely suspended or floating, and thus independent of water depth, at least during the active growing season.

Adaptive Plasticity

Individuals of many aquatic species are capable of assuming widely diverse forms in response to varying environmental factors in the habitats in which they find themselves. In the structure of stems and leaves especially, they can diverge surprisingly far from the average forms for their species in response to such factors as their state of maturity, the advancing season, as reflected by the length of the daylight period, the clarity, depth or temperature of the water (Cook 1968) or to the presence of competing vegetation. This plasticity enables a plant to exploit a complex of variable and constantly varying environments, such as the rising and falling interface between water and air, or seasonally fluctuating shores.

The plasticity of form displayed by aquatic plant species is in turn responsible for frequent problems in their recognition and classification. These problems must be approached with an open mind. The criteria by which differing character expressions are used to circumscribe and distinguish aquatic species must be accepted with a degree of flexibility that is seldom required when dealing with land plants. It is particularly important to keep in mind the possible ways in which ecological influences can affect the growth and morphology of an individual plant.

Adaptation of Pollination Systems

With increasing vegetative adaptation to life in water there is a progressive trend away from pollination systems dependent on insects to ones using the wind, or ultimately, water currents as the main pollinating agents. Associated with this trend is a reduction of the size of flowers. Many have become simple, inconspicuous and unisexual or even dioicous. They may be floating or even perpetually submerged; some extreme developments of the latter have thread-like, entangling pollen. Many of these flowers have the capacity for self-pollination in case the normal pollen-transfer vectors fail.

The objects of these modifications may be flowers barely recognizable to eyes accustomed to the showy flowers of insect-pollinated land plants.

Modifications of Fruit

Since seed-eating or fruit-eating land birds and wind may be ineffective as seed-dispersing agents on the water surface, aquatic plants have evolved fruits that can be dispersed by water. Some plants have fruits with spongy, buoyant tissues that enable them to float away from the parent plant. Others, especially those with many-seeded fruits, liberate the seeds through decay of the enclosing fruits under water.

Increased Reliance on Vegetative Means of Propagation

Many aquatic plants are notably shy flowerers, relying much more on vegetative means of propagation than most land plants. Some produce winter buds that are produced late in the growing season and are shed, to lie on the bottom of a water body over winter and germinate in spring to form new plants. Other plants produce extensive overwintering rhizomes or tubers, containing starch or other food reserves, for the same purpose. These bodies form important food sources for waterfowl and other water-inhabiting animals.

Evolution and Classification of the Group

It is generally agreed that this group includes, in the Superorder Alismatiflorae, the most primitive members of the Class Monocotyledonae. This is indicated by a number of features of structure and development in which they resemble primitive dicotyledons. Among these features are the occurrence of separate carpels in the flower, and the ununited margins of the carpels in some species. Even the great plasticity of form is itself a reflection of a lack of advanced specializations in the ancestral founding stock, while at the same time it is a character of positive value for facilitating adaptations to a variety of aquatic environments. The absence from most of these plants, except in the roots of some species, of the tracheae (tubular vessels) that characterize the water-conducting systems of most dicotyledons, may be a secondary reduction associated with an early phase of aquatic adaptation by the ancestors of this class.

Not every adaptive modification is found in any one species or family. It would appear that, given an initial trend toward the invasion of aquatic environments by an ancestral diversifying stock, the adaptive trends have been further developed by the diverging evolutionary branches independently of each other, to achieve parallel results. Thus we see an overall picture of a wide diversity of specializations developed independently in evolutionary lines of plants derived from a common, basically primitive stock of moisture-preferring or amphibious, but otherwise relatively unmodified ancestors. In this respect, it is interesting to point out that most modern students of plant evolution postulate that the monocotyledons, including this group of aquatic families, originated from an early, primitive stock of herbaceous, aquatic or amphibious dicotyledons whose surviving dicotyledonous descendants are the modern water-lilies (the Order Nymphaeales). The relationship between the Nymphaeales and primitive monocotyledons is indicated in many features, such as those of the stem anatomy (lack of cambium, scattered slender vascular strands, absence or near-absence of tracheae), the attachment of the ovules in the ovary, and the type of pollen (Takhtajan 1969; Cronquist 1988).

In particular, during this process of adaptation, the flowers of many of these plants have become so profoundly modified that the resolution of their structure, and even the delimitation of the individual flower from the inflorescence of which it is a part, is difficult, and for some plants, has been the subject of prolonged – and still unresolved – debate.

The varying interpretations of the structure and development of these plants, and particularly of their flowers and fruits, have given rise to a variety of conflicting hypotheses regarding the evolution of the group as a whole. Since concepts of evolution influence concepts of natural classification, these debates have led to a diversity of proposed systems of classification based on perceived relationships.

Such hypothetical natural systems of classification have undergone much evolutionary change in recent times. As our knowledge of plants has increased, progressively more use has been made of microscopic anatomy, including that of the pollen grains, internal tissues and cells, the chromosome complement, embryo development, and chemical contents and processes, in order to redefine groups of species, and rearrange them in progressively more genetically precise hierarchical classification systems.

Comprehensive treatments of modern classification systems are presented by Thorne 1976, Dahlgren, Clifford and Yeo 1985, and Cronquist 1988. Their treatments differ among themselves, depending on which characters each considered to be the most fundamentally important for distinguishing principal groups and subordinate groups within the Class Monocotyledonae.

Figure 1 (page 20) presents a very simplified scheme of the apparent evolutionary relationships among the families treated in this work, and of their placing in the overall classification of the monocotyledons, as they currently appear to me. In this figure, the families treated in this book are circled. The scheme presented combines features found in two classification systems. In general, it follows the system using subclasses presented by Cronquist (1988), but with some modifications derived from the work of Dahlgren, Clifford and Yeo (1985). In particular:

1. The Order Arales is in its own superorder, Ariflorae (after Dahlgren, Clifford and Yeo), allied in a common, larger and more inclusive complex with the Superorder Alismatiflorae. This would place the Arales in the Subclass Alismatidae of Cronquist.

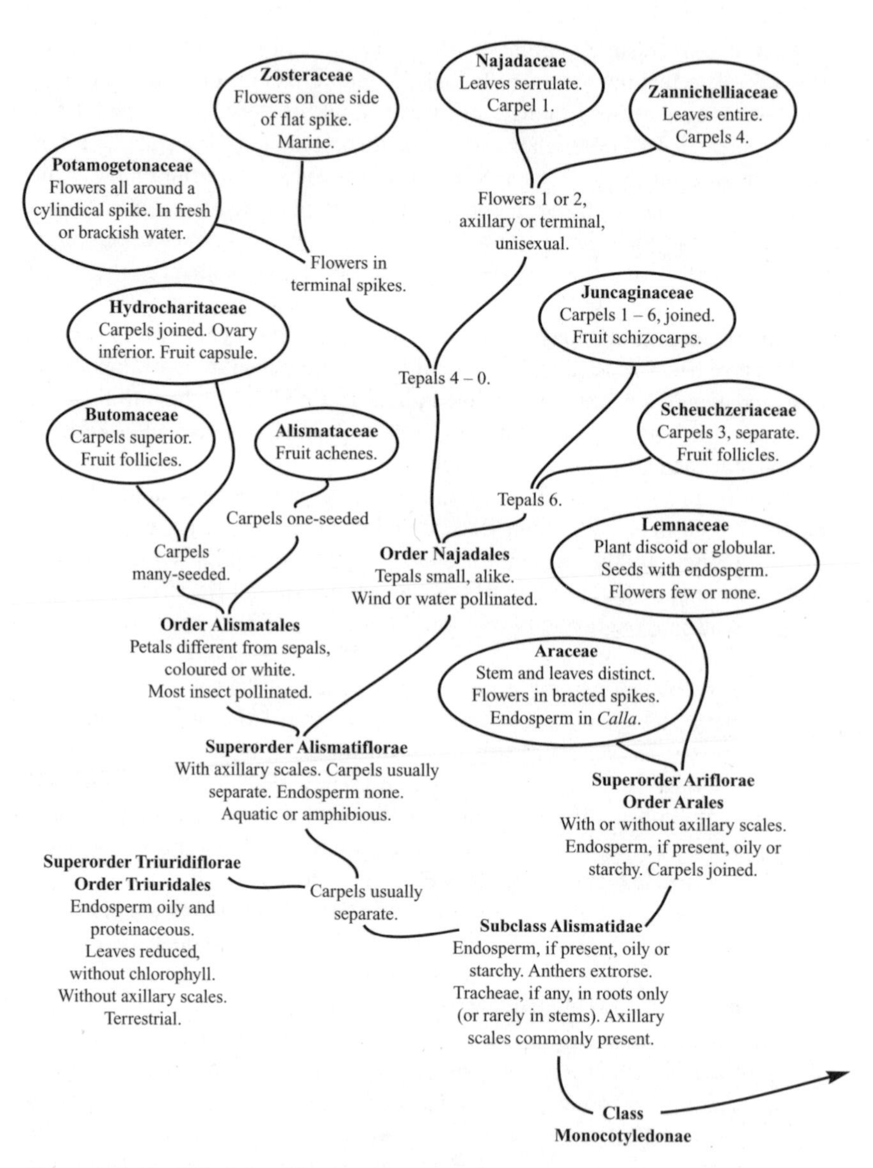

Figure 1. A simplified classification scheme of the monocotyledons, showing the positions of the families treated in this work (circled) in the overall classification of this class. Above: Subclass *Alismatidae.* Facing page: other subclasses.

---

2. The Family Pontederiaceae (of the Order Pontederiales) is placed in the Superorder Bromeliiflorae (as in Dahgren, Clifford and Yeo), on account of their starchy endosperm. This would place this family in the Subclass Zingiberidae, rather than in the Liliidae of Cronquist, which otherwise have oily and proteinaceous endosperm.

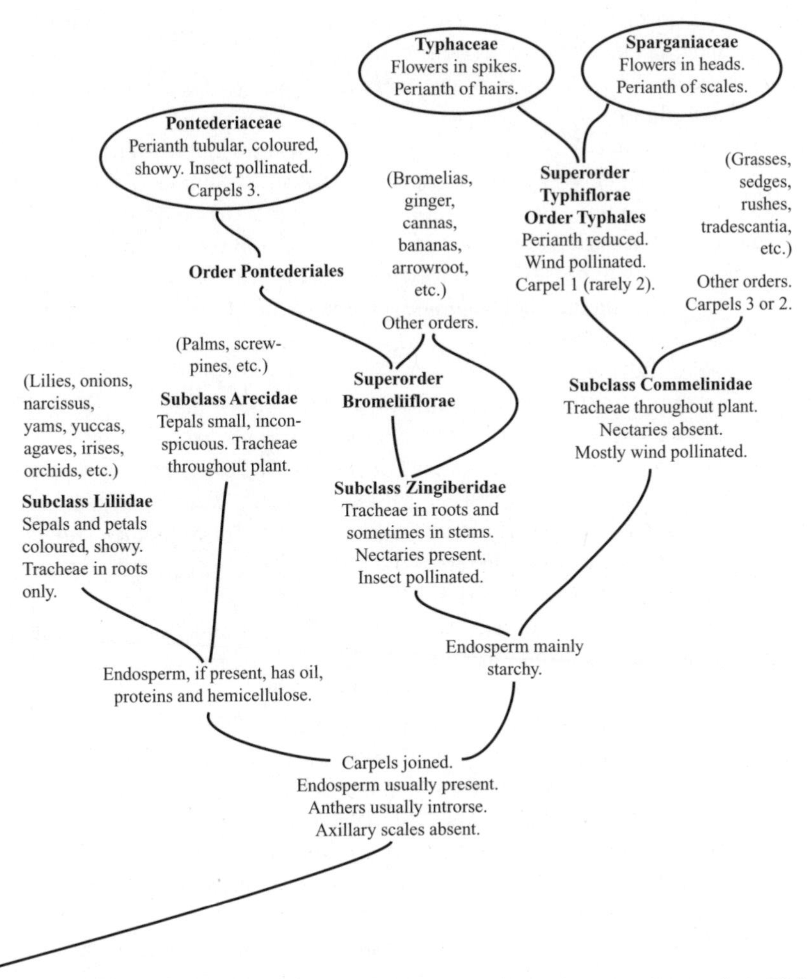

3. Typhales is in the Subclass Commelinidae (as done by Cronquist). This does not conflict with its place as a separate superorder, Typhiflorae, related to the Superorder Commeliniflorae by Thorne (1981), who also advocates combining the genera *Typha* and *Sparganium* in one family, the Typhaceae.
4. The arrangement of the orders and their families in the Superorder Alismatiflorae in the Subclass Alismatidae follows Dahlgren, Clifford and Yeo.

It can be seen that the families are arranged in a branching pattern, not in a linear sequence from primitive to advanced groups. This is a reflection of the ways in which various groups have either independently evolved different specialized features while retaining different primitive characteristics, or, in many cases, have independently evolved similar adaptations in parallel, in response to similar environmental influences. Thus, the sequence in which families are treated in this work (see the Checklist on pages 5 – 8) is only an arbitrary and approximate representation of evolutionary progress.

The superorders and orders whose members are described here are distinguished as follows:

1. The Superorder Alismatiflorae (Helobiae in the first edition) contains nine families of very diverse aspect in British Columbia. They display little that obviously relates them, but share several details of anatomy and development, such as the absence of endosperm in the mature seeds, the usually coiled or horseshoe-shaped embryo, the presence of minute scales (termed "squamulae intra-vaginales") in the axils of the leaves (and difficult to see), their usually separate carpels, and their universal preference for wet or fully aquatic habitats.

   Our nine families are divided between two orders. Plants in the Order Alismatales generally have white or coloured petals, and are adapted for pollination by insects. The genera *Elodea* and *Vallisneria* are exceptions to this rule, but show by anatomical and structural characters their relation to their family, the Hydrocharitaceae. The Order Najadales includes plants with minute greenish tepals or none, and are mostly adapted to pollination by wind or water currents.
2. Plants in the Superorder Ariflorae (Spadiciflorae of older classifications) and Order Arales usually have oily or starchy endosperm and a straight embryo, axillary scales in some, bundles (raphides) of calcium oxalate crystals in the cells of most species, and an ovary composed of carpels that are united if more than one per flower. In B.C., this group contains two families: the Araceae (arum) and the Lemnaceae (duckweeds).
3. The Superorder Bromeliiflorae, represented here by the Order Pontederiales (Farinosae in the first edition), have insect-pollinated flowers, with coloured sepals and petals united into a tubular perianth, and seeds with mainly starchy endosperm.
4. The Superorder Typhiflorae, consisting of the Order Typhales, has tracheae in all water-conducting tissues, in contrast to the other groups described here; it also has unisexual flowers in dense spikes or heads, the staminate spikes or heads above the pistillate ones on the same stem, scale-like or bristle-like tepals, a single carpel in each pistillate flower, and seeds with starchy endosperm. Both its families, the Typhaceae (cattails) and the Sparganiaceae (bur-reeds), are represented in British Columbia.

DIAGNOSTIC KEYS

The keys are intended as an aid in the identification of plants, using the most visible characters. They are not a system of natural classification, since the family relationships among the plants are often not evident from their gross appearance.

Many aquatic plants are such shy flowerers that they are seldom found bearing flowers or fruit. Thus it is most convenient to use readily visible features of vegetative form for diagnosing their identity, as far as possible.

The classification of plants, on the other hand, is determined by the characters of flowers, fruit, microscopic anatomy, chemical composition and development, which reflect the genetic parentage of the plants, but are often very difficult to observe.

Two general keys are presented on the following pages. The first is a key to all the families and genera containing aquatic plants known to occur in British Columbia, including pteridophytes and dicotyledons. The second key deals strictly with the families and genera covered in this work. These keys use gross vegetative features as far as possible. In many cases the keys lead directly to genera rather than to families, but in these cases the generic names are followed by the names, in brackets, of the families to which they belong.

In using a key, a succession of choices is made between paired, alternative character descriptions, starting with pair number one. At each stage, the character description is chosen that more nearly matches the specimen. Each successive choice leads to the name of a genus or family, or to the number of the next pair of alternatives to be chosen from. This process is repeated if necessary until the name of the plant group is reached. The description of the genus or family indicated should then be referred to for further details and keys to the species and perhaps varieties.

Key to the Families and Genera of Aquatic Plants in British Columbia

1a. Shrubs. 2
1b. Herbs..11
2a. Bud enclosed in a single bud scale. ....*Salix* (Salicaceae)
2b. Bud with two or more bud scales. 3
3a. Leaves opposite, entire. 4
3b. Leaves alternate. 5
4a. Leaves soft, deciduous. Tall shrubs, over a metre tall. ....*Cornus stolonifera* (Cornaceae)
4b. Leaves leathery, evergreen. Low shrubs, under a metre tall. ....*Kalmia* (Ericaceae)
5a. Leaf oblanceolate, toothed near apex, yellow-dotted, aromatic. ....*Myrica gale* (Myricaceae)
5b. Leaf wider: lanceolate to ovate or orbicular. 6

6a. Leaf conspicuously toothed or lobed...7
6b. Leaf entire..9
7a. Leaf sharply toothed or shallowly lobed. Flowers in terminal clusters, with white or pink petals. 8
7b. Leaf crenate, often shiny and leathery. Flowers in small lateral catkins, without petals..*Betula* (Betulaceae)
8a. Branchlets rather stout, with short stiff lateral spurs bearing leaves and terminal clusters of white flowers. Sometimes a small tree with shredding bark. ....*Malus diversifolia* (Rosaceae)
8b. Branchlets very slender, without short lateral stiff spurs, but with terminal dense panicles of small pink flowers. ....*Spiraea douglasii* (Rosaceae)
9a. Leaf scurfy with minute scales, flat, green beneath. ....*Chamaedaphne calyculata* (Ericaceae)
9b. Leaf not scurfy, with margins somewhat rolled under.......................10
10a. Young twigs and undersides of leaves whitened. ....*Andromeda polifolia* (Ericaceae)
10b. Young twigs and leaves not whitened.....................*Ledum* (Ericaceae)

11a. Leaves absent, not differentiated, or reduced to scales, cusps or basal sheaths. 12
11b. Leaves present, with well-developed blades and/or petioles. 17
12a. Minute free-floating plants, without distinction of stem and leaf. ....(Lemnaceae)
12b. Stems well-developed...13

13a. Stem bright yellow, threadlike, twining, parasite on *Salicornia*. ....*Cuscuta marina* (Cuscutaceae)

13b. Stem green or red, stout. 14

14a. Stem conspicuously jointed at regular intervals. Leaves scale-like or almost absent. 15

14b. Stem not jointed at regular intervals, with a few basal leaf sheaths only. 16

15a. Stem succulent, solid, each joint ending in a pair of opposite low cusps. Producing minute flowers in terminal succulent spikes. ....*Salicornia* (Chenopodiaceae)

15b. Stem parchment-textured, hollow, with whorls of joined dark scale-leaves. Producing spores in terminal cones with hexagonal sporophylls. ....*Equisetum* (Equisetaceae)

16a. Spike one, clearly terminal. ....*Eleocharis* (Cyperaceae)

16b. Spikes one to many, appearing lateral near top of stem, due to stem-like upright projecting bract. ....*Scirpus* (Cyperaceae)

17a. Leaf compound, or deeply dissected if simple. 18

17b. Leaf simple and unlobed. 34

18a. Leaves whorled or opposite. 19

18b. Leaves alternate. 22

19.a Leaf sectors forked, of cartilaginous texture when fresh. ....*Ceratophyllum* (Ceratophyllaceae)

19b. Leaf pinnately or palmately dissected, not of cartilaginous texture. .20

20a. Submerged leaves divided into at least three sectors at base. Emergent leaves, if any, broad and serrate. ....*Bidens* (Compositae)

20b. Leaves pinnately parted or lobed. 21

21a. Leaves mostly submerged, with filiform parts. ....*Myriophyllum* (Haloragidaceae)

21b. Leaves emergent, simple, flat, with lobed margins. ....*Lycopus* (Labiatae)

22a. Plant up to 5 cm long, normally floating, pinnately branched. Leaves minute, two-lobed. ....*Azolla* (Salviniaceae)

22b. Plant normally larger and rooted in mud. Leaves larger, with more than two lobes. 23

23a. Leaves palmately divided or having three leaflets. 24

23b. Leaves pinnately or bipinnately divided. 27

24a. Leaf with four fanlike leaflets. ....*Marsilea* (Marsileaceae)

24b. Leaf with odd number of leaflets, divisions or lobes. 25

25a. Leaf with three elliptic, entire leaflets. ..*Menyanthes* (Menyanthaceae)

25b. Leaf with three or more divisions that, if linear, are further divided or lobed. 26

26a. Submerged or floating plant with submerged leaves divided at least twice into linear divisions. Petiole base sheathing the stem. Separate white or yellow flowers. .... *Ranunculus* (Ranunculaceae)

26b. Tall erect plant with palmately lobed, flat leaves, normally all above the water. Petiole base and stipules not forming a sheath. Pink flowers in a terminal raceme..*Sidalcea* (Malvaceae)

27a. Plant with small bladders on leaves or stem. Leaf divisions linear or filiform. ....*Utricularia* (Lentibulariaceae)

27b. Plant without bladders on leaves or stem. Leaf divisions wider, flat. 28

28a. Leaf simply once pinnate..29

28b. Leaf bipinnate, with clearly defined leaflets. Petiole base sheathing stem. Flowers small white, in umbels..33

29a. Petiole base not sheathing stem. 30

29b. Petiole base sheathing stem. 31

30a. Petiole base with paired stipules. Leaflets irregular in size, sharply toothed. Erect plant. ....*Geum* (Rosaceae)

30b. Petiole base without stipules. Leaflets ovate to nearly circular, bluntly toothed to entire. Trailing or floating plant. ....*Rorippa nasturtium-aquaticum* (Cruciferae)

31a. Leaflets clearly distinct from the central axis of the leaf, lanceolate, 2.5 cm or more long..32

31b. Leaf variously lobed, but without clearly differentiated central axis and leaflets. Succulent plant of sea shores with creeping rhizome, and yellow buttonlike rayless flower heads. ....*Cotula coronopifolia* (Compositae)

32a. Erect plant, leaflets more than seven. ....*Sium suave* (Umbelliferae)

32b. Trailing and floating plant, leaflets five to seven. ....*Potentilla palustris* (Rosaceae)

33a. Leaf outline half to two-thirds as wide as its length. Lateral veins of leaflets running to notches in margin, or, if leaflets entire, with bulblets in their axils. Erect plants with chambered erect root stocks. ....*Cicuta* (Umbelliferae)

33b. Leaf outline broadly triangular, two-thirds to fully as wide as its length. Lateral veins of leaflets running into teeth. Plant with arching stems. ....*Oenanthe sarmentosa* (Umbelliferae)

34a. Leaves (except bracts) in a basal tuft, or on a basal rhizome. 35

34b. Leaves arising along stem (basal leaves may also be present). 64

35a. Leaves linear. 36

35b. Leaves with expanded blades. 49

36a. Leaf elongate, flat and ribbonlike, pliable. 37
36b. Leaf stiff, ascending or spreading, or, if pliable, not flat in cross-section. 40
37a. In fresh water. 38
37b. In sea water. .... Zosteraceae 39
38a. Leaves 4 – 5 mm wide, entire, relatively thick and tough, no rhizome. Flowers perfect, in a panicle. .... *Alisma gramineum* (Alismataceae)
38b. Leaves wider and thinner, minutely toothed, from a slender rhizome. Flowers unisexual: pistillate single on a long coiling peduncle, staminate many, minute, in a basal spathe. .... *Vallisneria* (Hydrocharitaceae)

39a. Rhizome short, irregular in thickness. Leaf narrow but thick and opaque. Dioicous plants with cordate fruits. Mostly on exposed, rocky coasts. .... *Phyllospadix*
39b. Rhizome relatively elongate, more uniform in thickness. Leaf wider, or, if narrow, thin and often translucent. Monoicous plants with ellipsoid fruit. Plants of sheltered bays and silty beaches. .... *Zostera*
40a. Plant with long slender stolons giving rise to leaves or leaf tufts at intervals. 41
40b. Plant with stout buried rhizome or corm, or annual. 42

41a. Leaf hollow, with externally visible partitions. Flowers minute, in umbels. .... *Lilaeopsis* (Umbelliferae)
41b. Leaf solid, unpartitioned. Flowers single or few, on stalks from nodes in stolons. .... *Ranunculus flammula* (Ranunculaceae)
42a. Leaf triangular in cross-section. Flowers pink in a large umbel. .... *Butomus umbellatus* (Butomaceae)
42b. Leaf not triangular in cross-section. 43
43a. Leaf flat and set edge-on to the flowering shoot, sheathing it at base. One leaflike bract at top of peduncle. Flower spike oblique. .... *Acorus* (Araceae)
43b. Leaf not set edge-on to shoot. 44

44a. Leaf bases widely distended around a corm, and enclosing spore-cases packed with white spores. Plants usually submerged. .... *Isoetes* (Isoetaceae)
44b. Leaf bases either not widely distended or not enclosing a corm. 45
45a. Leaf with distinct basal sheath 1 – 10 cm long. Plants of alkaline or saline marshes. 46
45b. Leaf not sheathing at the base or with a sheath not more than 5 mm long. 47

46a. Annual with terete leaf and one-carpelled, one-stamened flowers in a spike, with pistillate flowers in basal leaf sheaths. ....*Lilaea* (Juncaginaceae)

46b. Perennial with rhizomes and racemes of six-stamened, six-carpelled flowers. Leaf more-or-less semicircular in cross-section. ....*Triglochin* (Juncaginaceae)

47a. Leaf thick, hollow, rounded at tip, spreading. Peduncle several times as long as leaves, with white flowers in a raceme above water. Plant of shallow fresh water. ....*Lobelia dortmanna* (Lobeliaceae)

47b. Leaf slender, acute, ascending. 48

48a. Annual with flowers in short raceme a little taller than the leaves. Four sepals and four petals all separate. Plant of shallow fresh water. ....*Subularia* (Cruciferae)

48b. Perennial with single flowers on scapes shorter than leaves. Calyx cuplike; petals 5, joined. Plant of tidal marshes. ....*Limosella* (Scrophulariaceae)

49a. Leaf hollow, flasklike, with an erect "lid", often purple. Plant of bogs or fens. ....*Sarracenia* (Sarraceniaceae)

49b. Leaf of conventional form. 50

50a. Leaf margin toothed or crenate. 51

50b. Leaf margin entire. 54

51a. Underside of leaf white with fine cobwebby hair. Leaf arrowhead-shaped. ....*Petasites sagittatus* (Compositae)

51b. Leaf glabrous and green beneath, ovate to kidney-shaped or circular, crenate-margined. 52

52a. Plant with short ascending rhizome. Upper part of leaf sheath free from petiole, as a ligule. ....*Caltha* (Ranunculaceae)

52b. Plant with creeping rhizome or stolons. 53

53a. Rhizome stout. Sheath entirely joined to petiole, thus no ligule. Flowers white, in a raceme. ....*Nephrophyllidium* (Menyanthaceae)

53b. Rhizome or stolons threadlike. Flower single, pale violet to white. ....*Viola palustris* (Violaceae)

54a. Leaf blade with deep basal sinus. 55

54b. Leaf blade tapering, truncate or subcordate at base. 59

55a. Leaf apex acuminate or acute. 56

55b. Leaf apex rounded. Leaf often large, floating. 57

56a. Leaf blade arrowhead-shaped. Flowers in whorls, white. ....*Sagittaria* (Alismataceae)

56b. Leaf blade heart-shaped. Flowers white, in a dense spike (spadix) surrounded by a white spathe. ....*Calla palustris* (Araceae)

57a. Coarse plants, often with floating leaves 10 – 30 cm across, from a massive horizontal submerged rhizome. Flowers large. 58
57b. Smaller plant, up to 35 cm tall, inhabiting swamps. Leaves 1 – 4 cm across. Flower small, white......................... *Parnassia* (Saxifragaceae)
58a. Leaf margin rounded on either side of a deep basal sinus. Flower yellow. .... *Nuphar* (Nymphaeaceae)
58b. Leaf margin cuspidate on either side of basal sinus. Flower white or pink. .... *Nymphaea* (Nymphaeaceae)
59a. Leaf blade 40 – 150 cm long, conspicuously netted-veined. Flowers in a dense spadix surrounded by a yellow spathe. .... *Lysichiton americanum* (Araceae)
59b. Leaf blade much smaller, usually 1 – 10 cm long. 60

60a. Flowers in panicles with whorled branches....... *Alisma* (Alismataceae)
60b. Flowers single or few on a stem. 61
61a. Leaf almost sessile, obovate, slimy. Flower single, blue-purple, spurred. .... *Pinguicula* (Lentibulariaceae)
61b. Leaf petioled. 62
62a. Leaf blade ovate to circular, palmately veined. An ovate sessile bract part way up stem. Flower white, not spurred. .... *Parnassia* (Saxifragaceae)
62b. Leaf blade, if any, oblanceolate, three or more times as long as its width. 63
63a. Leaf blade tip rounded. Calyx cuplike. Petals joined, white to pink. .... *Limosella* (Scrophulariaceae)
63b. Leaf blade tip acute. Sepals and yellow petals all separate. Plant regularly stoloniferous. .... *Ranunculus flammula* (Ranunculaceae)
64a. Leaves alternate 65
64b. Leaves opposite, subopposite, or whorled. 93

65a. Leaves linear or narrowly lanceolate, often with no distinction between blade and petiole. 66
65b. At least some leaves with wide blades, either sessile or with distinct petioles. 86
66a. Leaves set edge-on to stem. 67
66b. Leaves flat side to stem, or cylindrical. 69
67a. Leaves 1 cm or more wide. Plant up to a metre or more tall. Flowers large, yellow. .... *Iris pseudacorus* (Iridaceae)
67b. Leaves less than 1 cm wide. Plant seldom over 30 cm tall. 68
68a. Flowers minute, crowded in dense brown heads. .... *Juncus ensifolius* (Juncaceae)
68b. Flowers few, on pedicels, yellow. .... *Sisyrinchium californicum* (Iridaceae)

69a. Stem triangular in cross-section, solid. Leaves three-ranked. Flowers minute, in scaly spikes. 70
69b. Stem not triangular in cross-section. 71
70a. Flowers of two kinds, staminate and pistillate, without perianth bristles. .... *Carex* (Cyperaceae)
70b. Flowers all bisexual, with perianth bristles. ....*Scirpus* (Cyperaceae)

71a. Leaf base not expanded into a sheath around the stem. Glabrous succulent plants of alkaline or saline marshes. Flowers minute, coloured as the foliage. ....*Suaeda* (Chenopodiaceae)
71b. Leaf base or petiole sheathing the stem. 72
72a. Stem hollow. Flowers two-ranked in small scaly spikes. 73
72b. Stem solid. 74

73a. Nodes noticeably swollen and solid. Internodes straight. Leaves two-ranked, with blades diverging from very long sheaths. Flowers in spikes in a terminal panicle .... Gramineae
73b. Nodes not swollen. Leaves clearly three-ranked. Flower spikes in axillary racemes. ....*Dulichium arundinaceum* (Cyperaceae)
74a. Plant erect. 75
74b. Plant with slender, flaccid, submerged stems. 80
75a. Flowers in dense globular heads, green to whitish. ....*Sparganium* (Sparganiaceae)
75b. Flowers in spikes, panicles, or asymmetric heads. 76

76a. Flowers in conspicuous long terminal spikes, the staminate spike above the pistillate one. Tall stout-stemmed plant with erect, pale green leaves. .... *Typha* (Typhaceae)
76b. Flowers all alike, perfect, in small spikes or other arrangements. Smaller plants with slender stems. 77
77a. Leaves with apical pores. Flowers in racemes. Fruits three follicles per flower..........................*Scheuchzeria palustris* (Scheuchzeriaceae)
77b. Leaves without apical pores. Flowers not in racemes........................ 78
78a. Stem terminating in one or more small scaly flower spikes. Fruits achenes with long perianth hairs, which collectively form a conspicuous cottony tuft. ....*Eriophorum* (Cyperaceae)
78b. Stem bearing clusters of small scaly flowers or flower spikes. Fruit without conspicuous cottony tufts of hairs. 79

79a. Flowers two or three in each of many tiny whitish spikes in one to three clusters. Perianth of 10 – 12 very short bristles. Plants of peat bogs. ....*Rhynchospora alba* (Cyperaceae)
79b. Flowers with six tepals of chaffy texture; brown or green, never white. ....*Juncus* (Juncaceae)

80a. Ligule absent, or if formed, less than 1 mm long. 81
80b. Ligule distinct, 2 mm or more long, or the sheath free from the leaf of petiole base, and thus all ligule..84
81a. Leaves filiform. Flowers with six chaffy-textured tepals. Fruit a capsule. ....*Juncus* (Juncaceae)
81b. Leaves ribbonlike, elongate. 82
82a. Plants of fresh water. Flowers in globose heads. ....*Sparganium* (Sparganiaceae)
82b. Plants of the sea or tidal lagoons. Flowers in flat spikes enclosed in sheathing leaf bases. 83

83a. Dioicous plants with cordate fruit. Rhizomes short, irregular in thickness. Leaf narrow but thick and opaque. On exposed rocky shores. ....*Phyllospadix* (Zosteraceae)
83b. Monoicous plants with ellipsoid fruit. Rhizomes more uniform in thickness, relatively elongate. Leaf wider, or if narrow, thin and often translucent. In sheltered bays and lagoons..........*Zostera* (Zosteraceae)
84a. Leaves narrowly lanceolate, tapering to base and apex, opaque, all alike, parallel-veined. Submerged. Flowers pale yellow, with six tepals and long perianth tube, emerging from sheath. ....*Zosterella* (Pontederiaceae)
84b. Leaves linear and parallel-veined, or if tapering, the submersed leaves translucent, with evident cross-veins between the parallel longitudinal veins. Flowers in a terminal spike; green, minute, with two or four tepals and no tube. ....Potamogetonaceae 85

85a. Flowers two per spike, with two stamens and two tepals. Achenes four, stipitate. Leaves filiform. Peduncle often long and coiled. ....*Ruppia* (Potamogetonaceae)
85b. Flowers more than two per spike, with four tepals and four stamens. Achenes sessile. Penduncles stouter, never coiled. Leaves diverse. ....*Potamogeton* (Potamogetonaceae)
86a. Leaf peltate, purple beneath. Submerged parts of plant covered with mucilage when young. Flowers solitary, dull purple. ....*Brasenia schreberi* (Nymphaeaceae)
86b. Leaf marginally attached, not peltate...87
87a. Sheath absent. Plant with short stiff appressed hairs. Flowers blue, small, in one-sided, uncoiling racemes..........*Myosotis* (Boraginaceae)
87b. Sheath present...88
88a. Sheath free from leaf or petiole base, or if attached, leaves of two kinds: floating leaves wider and thicker than submersed ones. ....*Potamogeton* (Potamogetonaceae)
88b. Sheath attached to leaf or petiole base, at least on basal leaves. Leaves all alike. 89

89a. Leaf netted-veined. 90
89b. Leaf parallel-veined from base or with closely parallel lateral veins from a midvein, entire. 91
90a. Leaf broadly elliptic, circular to kidney-shaped, palmately veined, crenate to toothed. Flowers single or few, not arranged in a spike. ....*Caltha* (Ranunculaceae)
90b. Leaf lanceolate to elliptic, or ovate, pinnately veined, entire. Flowers in a terminal spike, commonly pink. ....*Polygonum* (Polygonaceae)

91a. Leaf lanceolate, tapering to base and apex. Flowers one or two, pale yellow, with a perianth tube, emerging from a sheath. ....*Zosterella* (Pontederiaceae)
91b. Leaf ovate or broadly heart-shaped, cordate based. 92
92a. Leaf apex acuminate to attenuate. Flowers minute, white, in a spadix 2 – 4 cm long, subtended by a white spathe. ....*Calla palustris* (Araceae)
92b. Leaf apex rounded or blunt. Flowers blue, in a spike 6 – 15 cm long, subtended by a small green bract. ....*Pontederia cordata* (Pontederiaceae)

93a. Leaves in whorls of three to many. 94
93b. Leaves opposite. 97
94a. Leaves in whorls of four to many, narrow, entire. Plant submerged or the top often emergent. ....*Hippuris vulgaris* (Hippuridaceae)
94b. Leaves in whorls of three to six, finely serrulate-margined. 95
95a. Tall stout erect plants of wet ground, with broadly lanceolate, net-veined leaves. Flowers conspicuous, pink, spurred, very sweet-scented. ....*Impatiens glandulifera* (Balsaminaceae)
95b. Submerged, slender plants with small, one-veined, translucent leaves. Flowers white, floating. 96

96a. Leaves 20 – 35 mm long, in whorls of four to six. Flowers showy, with petals larger than sepals. ....*Egeria* (Hydrocharitaceae)
96b. Leaves 6 – 18 mm long, mostly in whorls of three. Flowers minute, with petals not larger than sepals. ....*Elodea* (Hydrocharitaceae)
97a. Leaves, though appearing opposite, really subopposite, the base and sheath of one in each pair arising within the sheath of the other. 98
97b. Leaves truly opposite. 99
98a. Leaf serrulate, continuous with its sheath. ....*Najas* (Najadaceae)
98b. Leaf entire, its sheath separate ....*Zannichellia* (Zannichelliaceae)

99a. Leaf margin toothed. 100
99b. Leaf margin entire. 108

100a. Flowers in terminal heads with yellow ray-flowers. Leaves pinnately many-veined, serrate or sometimes deeply dissected. ....*Bidens* (Compositae)
100b. Flowers not in terminal heads with yellow ray-flowers. 101
101a. Flowers one to three in a leaf axil. 102
101b. Flowers several in each axillary cluster. 104
102a. Flowers small, pale yellow to white. Calyx of five sepals. Corolla five-lobed at mouth. ....*Gratiola* (Scrophulariaceae)
102b. Flowers relatively large and showy. Calyx and corolla two-lipped. 103

103a. Flowers blue. Calyx with helmetlike crest on top. Stem square in cross-section. .... *Scutellaria galericulata* (Labiatae)
103b. Flowers yellow or red. Calyx without a crest on top. ....*Mimulus* (Scrophulariaceae)
104a. Flowers in racemes, usually blue. 105
104b. Flowers in dense axillary clusters, which may be congested together upward to form a thick spike. Stem square in cross-section. ....Labiatae 106
105a. Corolla wide open, four-lobed. Calyx without a dorsal crest. Fruit a two-lobed capsule. ....*Veronica* (Scrophulariaceae)
105b. Corolla tubular, two-lipped at mouth. Calyx with a dorsal crest. Fruit four nutlets per flower. ....*Scutellaria lateriflora* (Labiatae)

106a. Flower clusters absent from uppermost leaf axils. Plant soft-hairy, with a mintlike scent when crushed. ....*Mentha arvensis* (Labiatae)
106b. Flower clusters in middle and upper leaf axils. 107
107a. Foliage almost glabrous; the hairs visible only under magnification. Flowers minute, white. Plant scentless. ....*Lycopus* (Labiatae)
107b. Foliage distinctly rough-hairy. Flowers larger, mauve to pink or blue. Plant rank-smelling when crushed. ....*Stachys* (Labiatae)
108a. Plant with white stipules between the paired linear leaves. ....*Spergularia* (Caryophyllaceae)
108b. Stipules absent. 109
109a. Inflorescence or flowers terminal. 110
109b. Inflorescence or flowers axillary. 116
110a. Plants generally stiffly erect, not at all fleshy. 111
110b. Plants sprawling, often rooting at the nodes, with ascending branch ends, or if erect, fleshy. 114

111a. Plant less than 30 cm tall, usually several-stemmed. Leaves elliptic, 2 – 6 cm long. 112
111b. Plant 30 – 200 cm tall. Leaves lanceolate, 4 – 8 cm long. 113

112a. Flowers yellow, opening wide. Leaves and petals with black or clear dots. ....*Hypericum* (Hypericaceae)
112b. Flowers blue, the corolla funnel-shaped. Leaves and petals not dotted. ....*Gentiana sceptrum* (Gentianaceae)
113a. Leaves and flowers with black or purple dots. Flowers yellow. ....*Lysimachia terrestris* (Primulaceae)
113b. Leaves and flowers not dark-dotted. Flowers pink to purple. ....*Lythrum* (Lythraceae)

114a. Small plants with leaves 0.5 – 2 cm long. 115
114b. Larger, fleshy plants with leaves 2 – 4 cm long; leaf bases joined around the stem. Flowers in a terminal head. ....*Jaumea carnosa* (Compositae)
115a. Flower single, flesh-coloured to yellowish. Sepals five. Leaves and petals with black or clear dots. ....*Hypericum* (Hypericaceae)
115b. Flowers in terminal racemes, whitish. Leaves not dotted. ....*Montia fontana* (Portulaccaceae)
116a. Flowers in racemes. 117
116b. Flowers not in racemes. 118
117a. Leaves with dark dots. Stem erect. Flowers yellow, in racemes from lower leaf axils...........................*Lysimachia thyrsiflora* (Primulaceae)
117b. Leaves not dotted. Stem trailing with ascending branches. Flowers white, pink or blue..................................*Veronica* (Scrophulariaceae)

118a. Leaves sessile, at least if submerged, or without differentiation between petiole and blade...119
118b. Leaves with petioles. 124
119a. Leaf bases sheathing stem. Flowers minute, axillary, sessile to short-stalked. Minute plant with linear fleshy leaves. ....*Tilaea aquatica* (Crassulaceae)
119b. Leaf bases not sheathing stem. 120
120a. Flowers on pedicels 1 cm or more long. ...*Gratiola* (Scrophulariaceae)
120b. Flowers sessile or subsessile..121

121a. Leaves cordate-based. Erect emergent plant. Flowers minute, in a small cluster in each axil. Capsule globose. ....*Ammannia coccinea* (Lythraceae)
121b. Leaves not or scarcely cordate-based. 122
122a. Leaves membranous, at least the submerged ones linear-lanceolate with truncate or notched apices. Floating or terrestrial leaves, if present, petioled, spatulate, three-veined. Stem filiform, trailing. Flowers minute, green, usually one in an axil. ....*Callitriche* (Callitrichaceae)
122b. Leaves fleshy, elliptic. Stem relatively stout, trailing or erect. 123

123a. Leaves dotted (most visible in young leaves). Flowers in lower axils. Calyx cuplike, pink. Petals none. In saline or alkaline marshes. ....*Glaux maritima* (Primulaceae)

123b. Leaves not dotted. Flowers in upper axils or terminal. Calyx of free sepals, green. Petals small, whitish. On sea beaches. ....*Honkenya peploides* (Caryophyllaceae)

124a. Leaf blade ovate. Petiole fringed with white hairs. Flowers yellow, conspicuous, on axillary pedicels. ....*Lysimachia ciliata* (Primulaceae)

124b. Leaf blade elliptic to spatulate. Flowers sessile..............................125

125a. Erect emergent plant. Fruit a globose capsule. ....*Rotala ramosior* (Lythraceae)

125b. Trailing floating or matted plant..126

126a. Leaf blade acute at apex....................*Ludwigia palustris* (Onagraceae)

126b. Leaf blade rounded or notched at apex..127

127a. Leaf blade three-veined from base. Apex rounded or notched. Stipules none. Flower without petals, minute. Fruit a disclike schizocarp splitting into four one-seeded parts. ....*Callitriche* (Callitrichaceae)

127b. Leaf blade pinnately veined. Apex shallowly notched. Stipules present, minute. Flower with three minute petals. Fruit a globose capsule. ....*Elatine* (Elatinaceae)

Key to the Families and Genera Described in this Book

1a. Minute free-floating plants without distinct stems and leaves. ....Lemnaceae
1b. Plants with differentiated stems and leaves. 2
2a. Leaves all basal. 3
2b. Plants with leafy stems; often also with basal leaves. 13
3a. Mid-tidal to subtidal marine plants with flat, ribbonlike leaves. ....Zosteraceae
3b. Plants of fresh to brackish waters or marshes, or if in the sea, leaves not flat or ribbonlike. 4

4a. Leaves with broad, expanded blades. 5
4b. Leaves linear. 7
5a. Leaves heart-shaped, closely pinnately parallel-veined. ....*Calla* (Araceae)
5b. Leaves not heart-shaped, with conspicuous cross-veins, forming a net with the main veins. 6
6a. Leaves pinnately net-veined, very large, elliptic to broadly lanceolate. ....*Lysichiton* (Araceae)
6b. Leaves with spreading parallel veins and numerous oblique cross-veins. ....Alismataceae
7a. Leaves flat, ribbonlike. 8
7b. Leaves thick, triangular, circular or semicircular in cross-section. ...10

8a. Leaves arranged edge-on to stem. ....*Acorus* (Araceae)
8b. Leaves placed conventionally, flat side to stem. 9
9a. Leaves ribbonlike, with acute to rounded tips and scattered marginal teeth. Pistillate plant with single flowers on long coiling peduncles. Staminate plant with basal ensheathed clusters. ....*Vallisneria* (Hydrocharitaceae)
9b. Leaves gradually tapering to acute tips, or if ribbonlike, without marginal teeth. Flowers in whorls or panicles. ....Alismataceae

10a. Leaf triangular in cross-section, erect. Flowers in a false umbel, pink, showy. ....*Butomus* (Butomaceae)
10b. Leaf circular or semicircular in cross-section. Flowers in racemes or spikes. 11
11a. Leaf circular in cross-section. Flowers in spikes. ....*Lilaea* (Juncaginaceae)
11b. Leaf semicircular in section. Flowers in racemes. 12

12a. Leaf with apical pore. Raceme with bracts and up to five flowers. ....*Scheuchzeria* (Scheuchzeriaceae)

12b. Leaf without apical pore. Raceme bractless, with eight to many flowers. ....*Triglochin* (Juncaginaceae)

13a. Leaf with upstanding heart-shaped blade, with closely parallel lateral veins...14

13b. Leaf linear, or, if with an expanded blade, the blade floating or submersed. 15

14a. Visible stem erect, with a leaf or two and a spike of blue flowers. ....*Pontederia* (Pontederiaceae)

14b. Visible stem trailing, with ascending leaves and white spathe and spadix. ....*Calla* (Araceae)

15a. Flowers in dense globular heads. Leaves linear. ....*Sparganium* (Sparganiaceae)

15b. Flowers not in heads. Leaves various. 16

16a. Leaves erect, straplike. Flowers in conspicuous spikes. ....*Typha* (Typhaceae)

16b. Leaves floating or submersed. 17

17a. Leaves subopposite or whorled..18

17b. Leaves mainly alternate. 20

18a. Leaves in whorls of three or four, the whorls often crowded. ....*Egeria* or *Elodea* (Hydrocharitaceae)

18b. Leaves subopposite, with minute flowers or fruit at their bases........19

19a. Leaves oblong-lanceolate, serrulate, their dilated basal sheaths not funnel-like or surrounding the stem. ....*Najas* (Najadaceae)

19b. Leaves linear, entire, with separate, funnel-like sheaths surrounding the stem when young. ....*Zannichellia* (Zannichelliaceae)

20a. Plants of marine or intertidal habitats, with elongate, ribbonlike leaves. ....Zosteraceae

20b. Plants of freshwater habitats, or if marine or intertidal, leaves not ribbonlike...21

21a. Leaves slenderly lanceolate, tapered to bases and apices, adherent at bases to ligular sheaths. Flowers one or two in axils, complete, pale yellow, with long perianth tubes. ....*Zosterella* (Pontederiaceae)

21b. At least the submersed leaves either linear to threadlike, with untapered bases adherent to sheaths, or free from sheaths, linear and ribbonlike, lanceolate, ovate or elliptic. Flowers minute, in small terminal spikes, greenish, without perianth tube. ....Potamogetonaceae

DESCRIPTIONS OF FAMILIES, GENERA AND SPECIES

The Family Butomaceae — Flowering-rush Family

Perennial, rhizomatous, rushlike herbs with two-ranked basal linear leaves, triangular in cross-section above, without distinction of blades and petioles; the sheathing bases with axillary buds and groups of narrow axillary scales. The leaves are erect, emergent from the water, and often twisted.

The inflorescence terminates a tall, axillary, smooth scape. It is umbel-like in form, but is developmentally a very condensed cymose panicle, based on a fundamentally alternate branching habit, in which pairs of successive internodes fail to elongate. Thus the peduncle is topped by a false whorl of three bracts, each subtending one or more axillary branches.

The terminal flower, and the terminal flower of each branch, open first, and are followed by others formed in succession from branchlet buds in the axils of bractlike prophylls at the bases of the branches (Singh and Satler 1974). Thus the flowers open in succession and the false "umbel" is repeatedly replenished by the development of younger flowers through at least the early summer (Charlton and Ahmed 1973).

Each flower has three sepals and three larger petals, all coloured a pale pink to purplish, and persistent into the fruiting stage. There are commonly nine stamens and several (commonly six) carpels in a whorl. The carpels are separate at first, but become joined at their bases by the fruiting stage through the upgrowth of a ring of underlying tissue. Each carpel tapers into a distinct style, which bears the stigmas on the upper ventral margins. In each flower, the pollen is shed before the stigmas become receptive. The numerous ovules are scattered over the inner surface of the carpels. The ventral margins of the carpels are not joined, as in most flowering plants, but are merely pressed together.

The fruits are whorls of follicles containing many seeds, which are liberated by separation of the ventral carpel margins. The mature seeds are longitudinally ribbed, and contain straight embryos and no endosperm. $2n = 26, 39$. Figure 2.

This family, as currently accepted, consists of one genus and one species. Five other, mostly tropical, genera, formerly included in this family, are treated by modern systematists as a separate family: the Limnocharitaceae.

Figure 2. *Butomus umbellatus*: A, inflorescence and leaf apex; B, rhizome and shoot base; C, flower; D, fruit (cluster of follicles); E, seeds.

The Genus *Butomus* L.

The only genus, consisting of one species.

Butomus umbellatus L. — Flowering-rush

The only species, with characteristics of the family (see figure 2). Native to temperate Eurasia, this species has been introduced into North America. It is grown as an ornamental plant for aquatic or marshy sites because of its showy, attractive flowers. It is now widespread on this continent; but in British Columbia, it has been collected only at Hatzic Lake, near Mission, where it forms small colonies near the shore in water 40 – 70 cm deep.

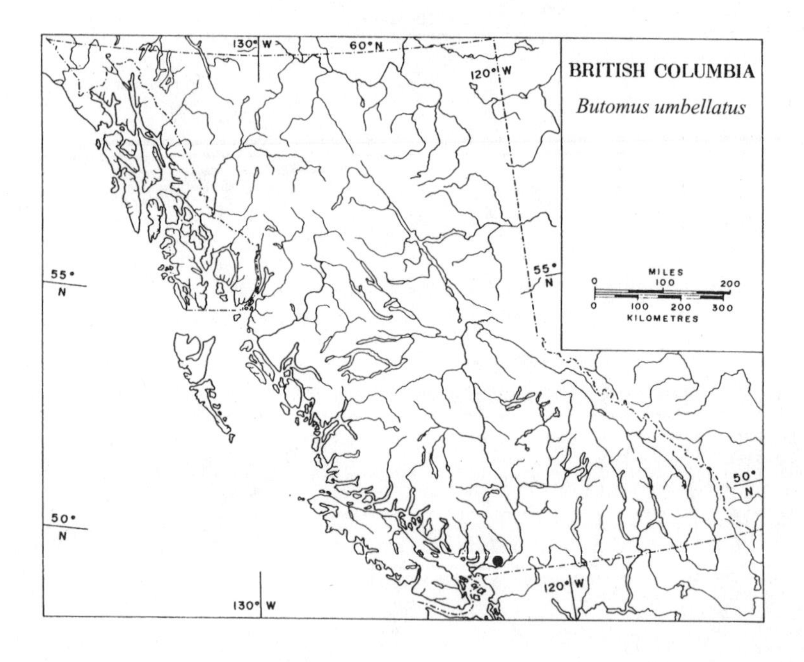

The Family Hydrocharitaceae **Frog-bit Family**

Mostly dioicous aquatic herbs, all of ours in fresh waters, with diverse vegetative forms. Flowers one or more arising in sessile or pedunculate axillary spathes, each of which consists of one or more joined bracts. Sepals and petals three each, or the petals lacking; the bases of the perianth parts often united into a hypanthium that may be tubular and elongate. Staminate flower has two to twelve stamens. Pistillate flower has an inferior three-carpelled (in ours) but unilocular ovary and bilobed stigmas. Ovary matures into an indehiscent, many-seeded fruit. Seed has a straight embryo and usually no endosperm.

Key to Genera in British Columbia

1a. Stem permanently rooted in bottom sediment, with long, ribbonlike basal leaves. .... *Vallisneria*
1b. Plant becoming suspended in water. Stem with regularly placed whorls of small ovate or lanceolate leaves 2
2a. Leaves in whorls of mostly four to six, 20 – 35 mm long. Petals much larger than sepals. .... *Egeria*
2b. Leaves in whorls mostly of three, 6 – 18 mm long. Petals not larger than sepals. .... *Elodea*

The Genus *Egeria* Planchon

Rooted or suspended dioicous perennials, with the aspect of *Elodea*, the slender branching stems with regularly spaced whorls of four to six (or, rarely, three) leaves. Leaves small, sessile, lanceolate, serrulate-margined, veinless except for the midvein, mostly two cells thick, consisting of the upper and lower epidermal layers.

Both pistillate and staminate flowers are permanently attached to the plant, and sessile in axillary spathes that are minutely notched at apex and open to base on one side. Staminate flowers two to several in a spathe, expanding in succession by elongation of their threadlike hypanthia. Sepals three, reflexed; petals three, much larger than the sepals; stamens nine, separate. Pollen transfer is carried out by flies.

Egeria is sometimes (as in the first edition of this work) included in *Elodia*.

Egeria densa Planchon **Large-flowered Water Weed**
(*Elodea densa* (Planchon) Caspary)

Resembling the more common *Elodea canadensis*, but larger, more sparingly branched, and more densely leafy, with closely spaced whorls. Leaves 10 – 35 mm long by 2 – 3 mm wide, tapering from near the base to an acute to acuminate apex, and serrulate on the margins.

Flowers showy. Sepals broadly ovate to obovate, 3 – 5 mm long by 2 mm wide, greenish, reflexed. Petals obovate to suborbicular, 7 – 10 mm long, white, spreading and ascending, with a slightly crumpled texture. Stamens nine, separate, the six outer ones spreading, the three inner ones more or less erect. Ovary none in ours. The hypanthium elongates up to about 7 cm long to bring the floral appendages above the water surface, on which the flower floats on the tips of its reflexed, water repellent sepal. 2n = 48. Figure 3.

Native to South America, this species has been introduced extensively in other lands, and has become a pest in some warmer countries. It is cultivated in aquariums and garden ponds, from which it may be "liberated" into nearby natural waters, where it can maintain itself and spread by fragmentation. Only staminate plants are known in North America, reflecting the nature of the originally introduced plant. In British Columbia, this species has been found in Florence and Glen lakes, and reported from Langford Lake, all in close proximity near Victoria (no map).

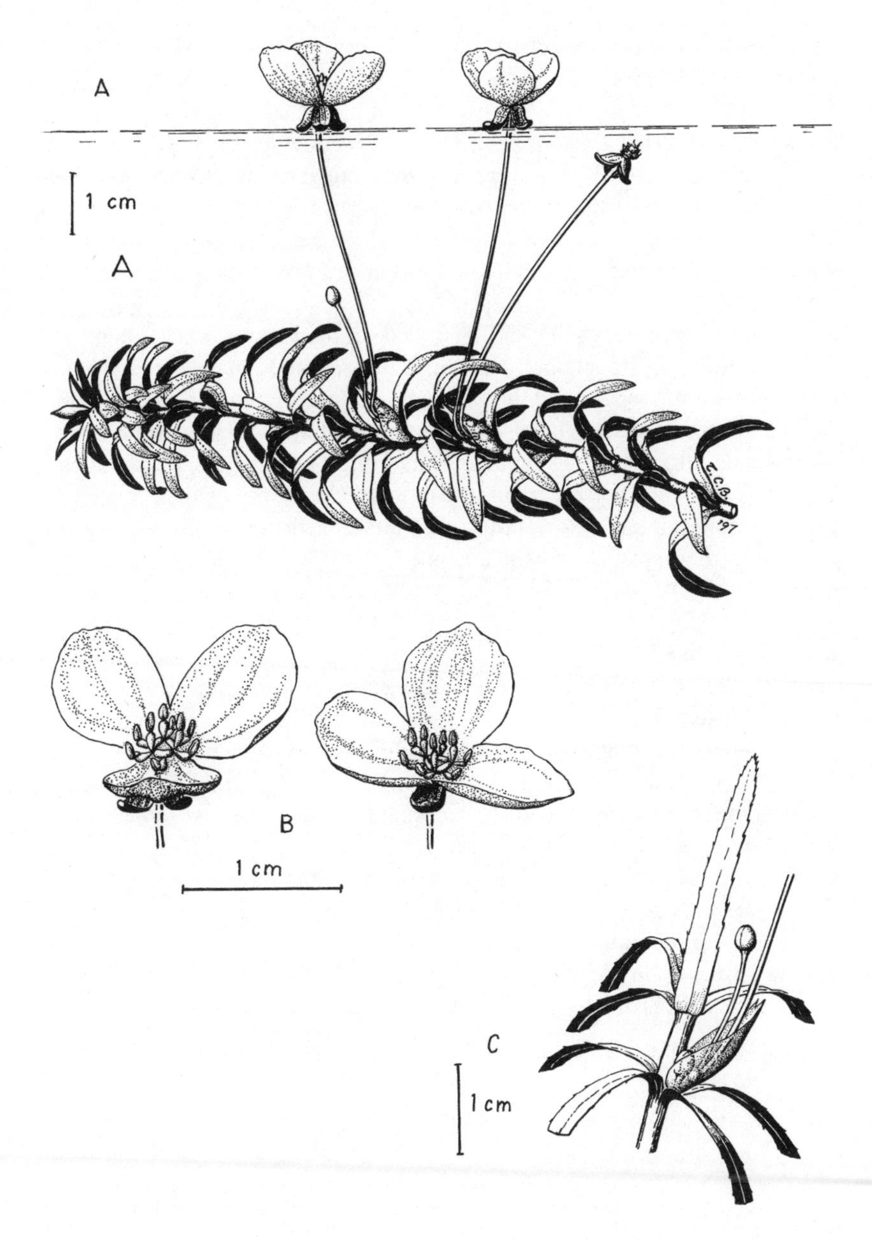

Figure 3. *Egeria densa*: A, staminate flowering shoot, B, staminate flowers; C, leaf, leaf bases and spathe, with buds and hypanthium base of open flower.

The Genus *Elodea* Michaux (*Anacharis* Richard) — Water Weed

Rooted or suspended dioicous perennials with sessile leaves mostly in whorls of three or in opposite pairs, regularly distributed along slender branching stems. Leaves small, ovate to lanceolate, veinless except for the midvein; away from the vein, consisting of the upper and lower epidermal layers. Axillary scales in pairs. Scattered long shining roots arise from occasional nodes along the stem.

Flowers sessile in sessile, axillary, two-toothed spathes. The staminate flower solitary in its spathe, and consisting of three sepals, three petals narrower than the sepals, and three to nine stamens, the innermost three with coherent filament bases. The pistillate flower solitary, with three sepals, three or no petals, three slender staminodes, and three notched or forked stigmas. Open flowers reach the water surface through elongation of their hypanthia, except for the staminate flower of *Elodea nuttallii*; the ovaries remaining sessile within their spathes. This genus is adapted for pollination by wind and water.

Key to Species

1a. Leaves ovate or broadly lanceolate, entire or with a few teeth near the obtuse to rounded apex. Staminate flower always attached and apparently long-stalked by extension of the hypanthium. .... *E. canadensis*

1b. Leaves narrowly lanceolate, tapering from near the base to the acute apex, serrulate along the sides. 2

2a. Leaves up to 15 mm long by 2 mm wide. Staminate flowers detached and floating free for pollen release. Pistillate flowers with sepals 1 – 1.8 mm long. .... *E. nuttallii*

2b. Leaves up to 25 mm long by 2.5 mm wide. Staminate flowers always attached. Pistillate flowers with sepals 3 – 3.5 mm long. .... *E. callitrichoides*

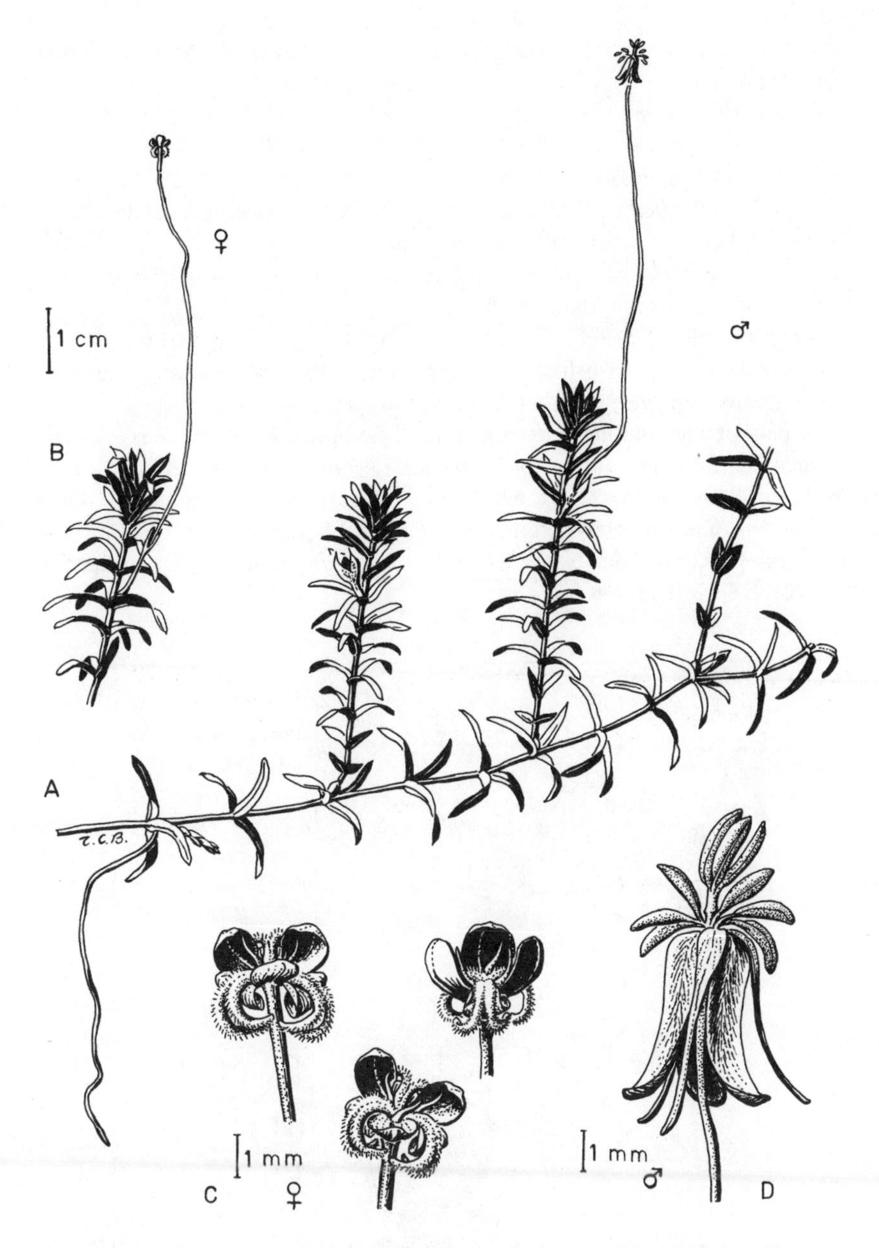

Figure 4. *Elodea canadensis*: A, staminate flowering shoot and flower; B, pistillate flowering shoot; C, pistillate flowers (upper parts); D, staminate flower.

Elodea canadensis Richard *in* Michaux — Canada Water Weed

***Anacharis canadensis* (Richard) Planchon**

Stems bearing opposite leaves at first, but later, mainly leaves in whorls of three, often crowded and recurving. Leaves ovate to broadly lanceolate, obtuse or rounded at apex, entire or with a few minute teeth near the apex, 5 – 15 mm by 1 – 4 mm wide. Mesophyll tissue is practically absent from the leaf, and photosynthesis is carried out by the epidermis.

Flowers white, the floral appendages of both staminate and pistillate flowers floating on the water surface, and connected to their sessile floral bases by their greatly elongating slender tubular hypanthia.

Sepals of the staminate flower narrowly elliptic, 4 – 5 mm long; petals lanceolate, as long as the sepals or slightly longer; both sepals and petals reflexed. Stamens nine, the three or more inner ones on a common staminal column or tube. The non-wetting pollen grains, liberated by explosive rupture of the anthers and blown by the wind, float across the water surface to the stigmas of the pistillate flowers.

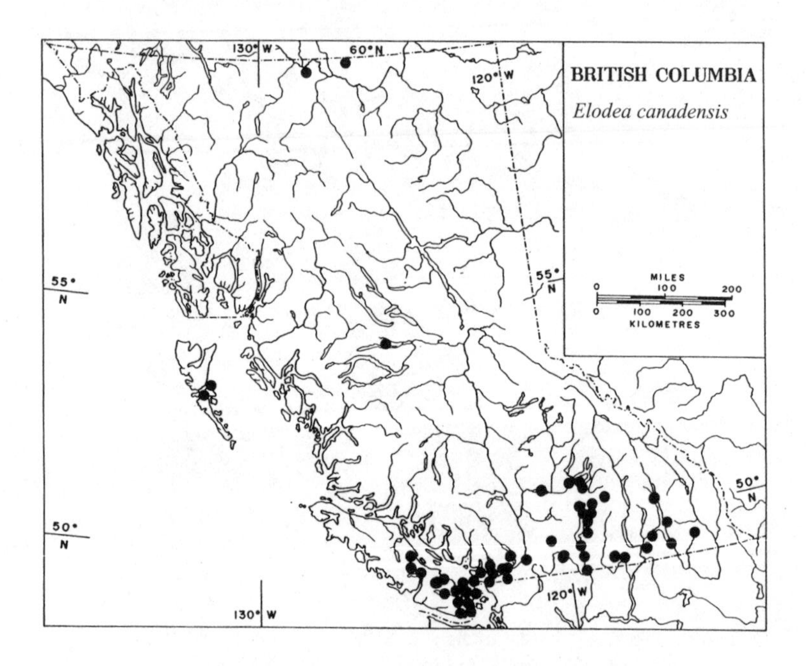

The pistillate flower is smaller than the staminate flower, with a threadlike hypanthium up to 30 cm long, raising the floral appendages to the water surface; with almost equal reflexed sepals and spreading or ascending petals 2 – 2.5 mm long; three hornlike, antherless staminodes 0.5 – 1.5 mm long, and bifurcating white stigmas whose tips are recurved between the petals to make contact with the water-borne pollen. Fruiting capsule 6 mm long, with seeds 4 – 4.5 mm long. 2n = 24, 48. Figure 4.

This widespread North American aquatic weed is now also well established in the Old World. Abundant across southern British Columbia, it is widely distributed but seldom collected further north. It reaches within a few kilometres of the border with the Northwest Territiories; though it is not recorded from the Territories by Porsild and Cody (1980).

It becomes particularly abundant in some polluted lakes, where it forms extensive floating and suspended mats. The plant can occur in shallow water or in water up to 3 metres deep. Commonly attached to the bottom early in the season, it later often becomes detached from the bottom, and suspended at or just beneath the water surface.

Flowering occurs between July and September, large areas often flowering at once. But the flowers are easily overlooked and seldom collected, because of their small size and their single and scattered distribution over the water's surface. Vegetative perennation occurs through the production of detachable overwintering shoot apices with short internodes enclosed in densely overlapping foliage.

Elodea callitrichoides (Richard) Caspary — South American Water Weed

***Elodea ernstiae* St John**

Elodea callitrichoides vegetatively resembles a rather coarse *E. nuttallii*. The leaves in whorls of three, or the lowest ones on a shoot opposite; up to 25 mm long by 2.5 mm wide, the margins serrulate with teeth projecting 0.1 – 0.15 mm.

Flowers larger than those of *E. nuttallii*. The staminate flowers remain attached to the plant. The pistillate flowers have sepals 3 – 3.5 mm long (not illustrated). 2n = 32.

This species, a native of South America, has been introduced into North America and Europe as an aquarium plant.

One specimen, collected at Still Creek in Burnaby, and now in the collection at the University of British Columbia, at first thought to be *E. nuttallii* from its vegetative appearance, has been annotated by Paul Catling as being *E. ernstiae* on the basis of a bud about 3 mm long. If this bud is that of a flower, then *E. callitrichoides* (or *E. ernstiae*) must be its correct identity.

I have not seen any other specimens that possess flowers from this "*E. nuttallii*" population between Vancouver and Pitt Lake. It would be most desirable to seek flowering material of this population to confirm its identity. The possibility exists that it could be found to be an invading population of yet another exotic weed.

Elodea nuttallii (Planchon) St John — Nuttall's Water Weed

***Anacharis nuttallii* Planchon**

Similar in form to *Elodea canadensis*, but tending more to remain rooted in the bottom sediments. Leaves apparently less crowded, in whorls of three or occasionally four, or opposite at the lowest nodes of a shoot, narrowly lanceolate, 6 – 13 (– 20) mm long by up to 1.5 (– 2) mm wide, tapering from near the base to an acute apex, the margin serrulate with teeth projecting 0.05 – 0.1 mm. Leaves on flowering stems noticeably smaller and more abruptly tapering than those on sterile shoots.

Staminate spathe sessile, globular, 2 – 3 mm wide, with two projecting points, containing one flower. The flower bud is initially sessile in its spathe, but at pollen maturity the bud breaks free from its spathe, which opens at least half way to the base on one side to release it. The bud floats to the water surface, where it opens and drifts by wind to a pistillate flower; the anthers open and the non-wetting pollen may be shed onto the water surface. Sepals reflexed, about 2 mm long, white or marked with purplish, the petals smaller or absent. Stamens nine, their filament bases joined into a staminal column, three of the stamens above the other six.

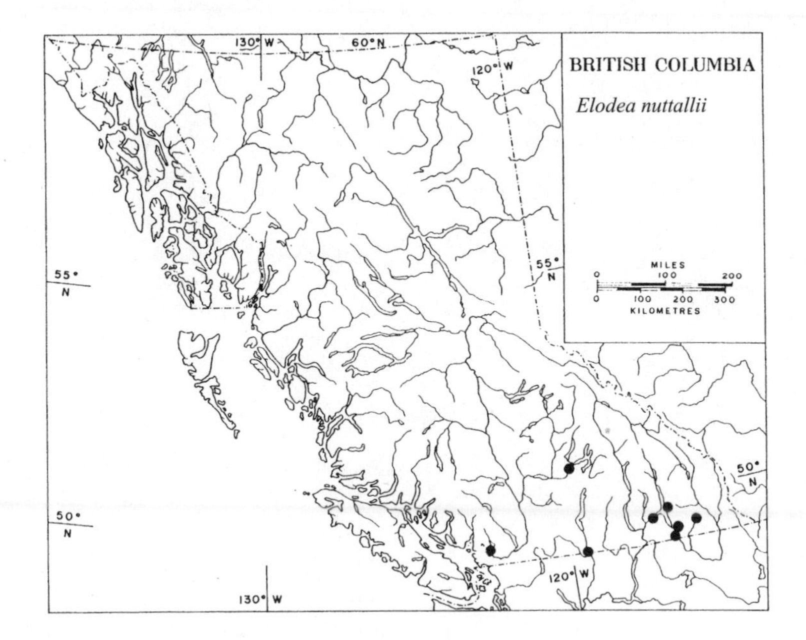

Pistillate flower sessile in a tubular spathe with an oblique orifice, the ovary, containing five or six ovules, remaining in the bottom of the spathe while the hypanthium elongates up to 10 cm. Sepals and petals spreading, about 1.5 mm long. Three antherless staminodes are about 0.5 mm long; stigmas three, spreading, shallowly bilobed or notched at their tips, purplish. Fruit a capsule 5 – 7 mm long, its seed 4 mm long. 2n = 48. Figures 5 and 6.

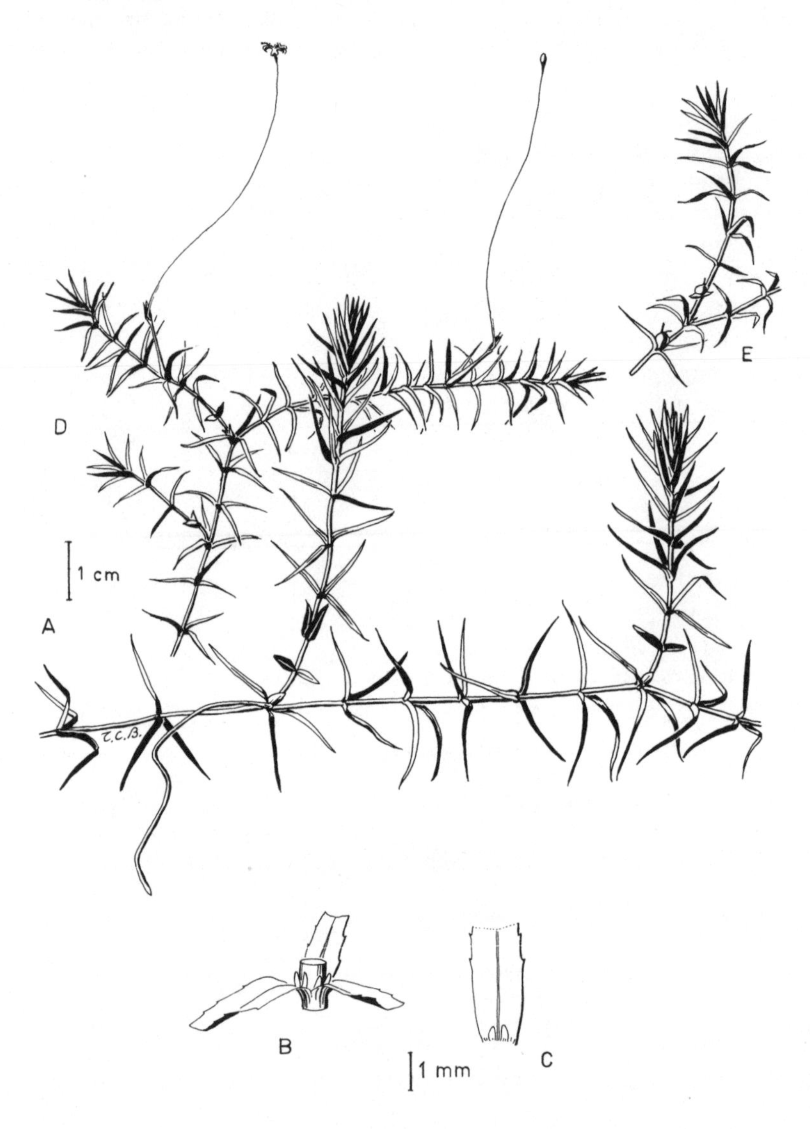

Figure 5. *Elodea nuttallii*: A, sterile shoot; B, node of stem, leaf bases and axillary scales; C, leaf base, showing axillary scales; D, pistillate flowering shoot; E, staminate shoot with two spathes.

Widespread in central and eastern North America; now naturalized and spreading in much of Europe. In British Columbia, it is found mainly in the southern Kootenay region, where it is uncommon in shallow water with clay or silty bottom in lakes or the backwaters of rivers. Flowers are produced in September, and are very seldom collected. An outlying population resembling *E. nuttallii* in the lower Fraser River valley is currently of uncertain identity, since no flowering material has yet been seen. It may possibly be *E. callitrichoides* (see note under *E. callitrichoides*).

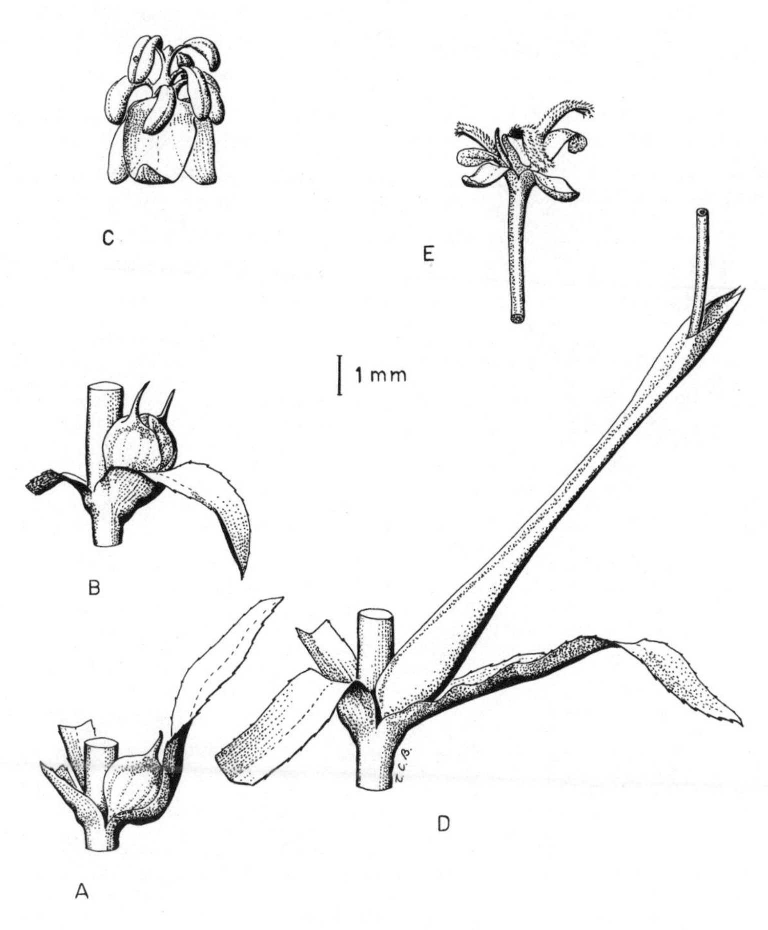

Figure 6. *Elodea nuttallii* (spathes and flowers): A, staminate spathe (closed); B, staminate spathe (open), the enclosed flower bud visible; C, staminate flower; D, pistillate spathe with emerging flower hypanthium; E, upper, floating part of pistillate flower.

The Genus *Vallisneria* L. **Tapegrass**

Submersed rhizomatous perennials, with short leafy shoots terminating rhizome segments; extensions of the rhizome growing from buds in the axils of reduced or scale-like basal leaves. Leaves tufted, ribbonlike, their length influenced by water depth.

Plants dioicous. Flowers arising within a spathe on the end of a peduncle, which is short in staminate plants, and greatly elongated in pistillate plants.

Staminate flowers minute, short-pedicelled at inception, with buds densely aggregated on a spadix enclosed in a submerged spathe. Each flower consists of three sepals, one to three petals, and two stamens. Pistillate flower solitary in a two-lobed sheathing spathe, and consisting of an elongate inferior ovary crowned by the three sepals, three petals and three forked, spreading, shortly villous stigmas.

When the pollen is ripe, the staminate spathe ruptures and the minute flower buds, buoyant by enclosed gas bubbles, break loose and float up to the water surface. There, they burst open (see figure 7C), then, floating on their water-repellent sepals, they drift with the wind. The pistillate flower, in its spathe, rises to the surface on a long, flexible penduncle (figure 8A) and floats there with its sepals, petals and stigmas exposed (Arber 1920, Svedelius 1932).

The relatively massive pistillate flower, floating on its water-repellent sepals, causes a local depression of the water surface. Near enough is good enough. Any staminate flowers drifting near a pistillate flower will slide down the sloping meniscus to collide with the exposed stigmas. Several staminate flowers may be drawn in to cluster around the pistillate flower and effect pollination. After pollination, the pistillate flower's peduncle bends or coils helically to draw the fruit deep beneath the surface for ripening.

Because of their rhizomatous structure, tapegrasses can spread vegetatively and establish extensive colonies whose members are all clones, genetically identical layered branches of the original seedling plant.

Key to Species

(based on Svedelius 1932)

1a. Pistillate peduncles helically coiled, 30 – 60 cm long, sepals 1 – 4 mm long, ovary straight. Staminate flower with widely divergent stamens. .... *V. spiralis*

1b. Pistillate peduncle curved but not coiled, 5 – 10 cm long, sepals 5 – 6 mm long, ovary curved. Staminate flower with stamens parallel and partially or wholly united. .... *V. americana*

Vallisneria americana Michaux — American Tapegrass

Leaves ribbonlike but tapered toward the ends to rounded or acute apices, usually minutely toothed, especially near the tips; their lengths varying up to 2 metres long, according to the depth of the water, the tips floating or submerged; the width 5 – 20 mm, with three or more longitudinal veins and scattered cross-veins.

Staminate spathe deeply submerged, 1 – 2 cm long, tapered at base into a short stout peduncle up to 5 cm long by 1.5 – 3 mm thick, or nearly sessile; each spathe containing hundreds of crowded, minute white flower buds on short pedicels. The staminate flower, when open, is 1 – 1.5 mm long, with three unequal sepals about 0.5 – 0.8 mm long, one minute petal, a pair of obliquely upstanding fertile stamens with their filaments nearly parallel and partially united, bearing exposed pollen masses of relatively few grains, and a minute staminode. Figure 7.

The shiny pistillate peduncle becomes greatly elongated and curved to allow the solitary flower to float on the water surface. The spathe, 2 – 3 cm long and two-lobed at the apex, loosely surrounds the ovary of the flower. Pistillate flower whitish, with three sepals 5 – 6 mm long, three minute petals, three minute staminodes, and three conspicuous, forked stigmas. The flower floats, at first vertically, later often obliquely to the surface or nearly horizontally, with the sepals resting on the surface.

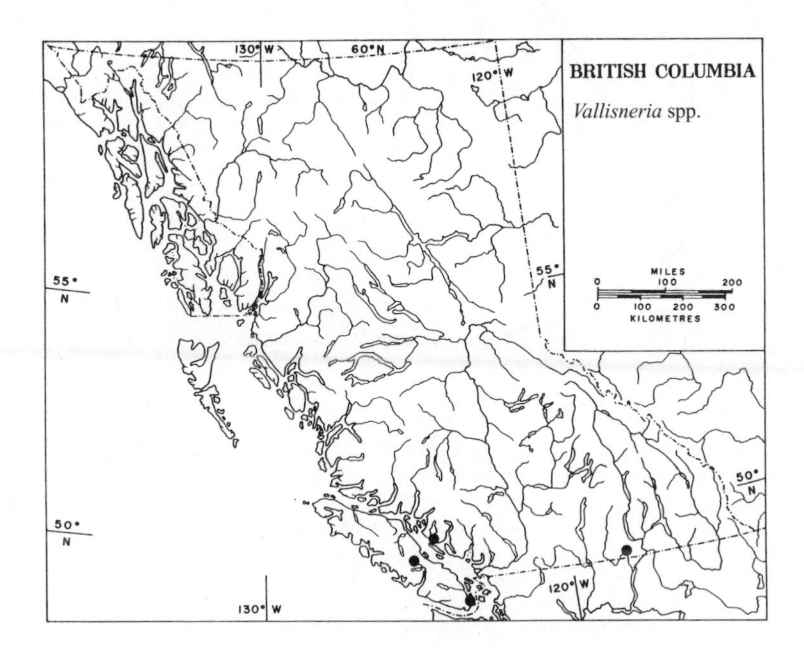

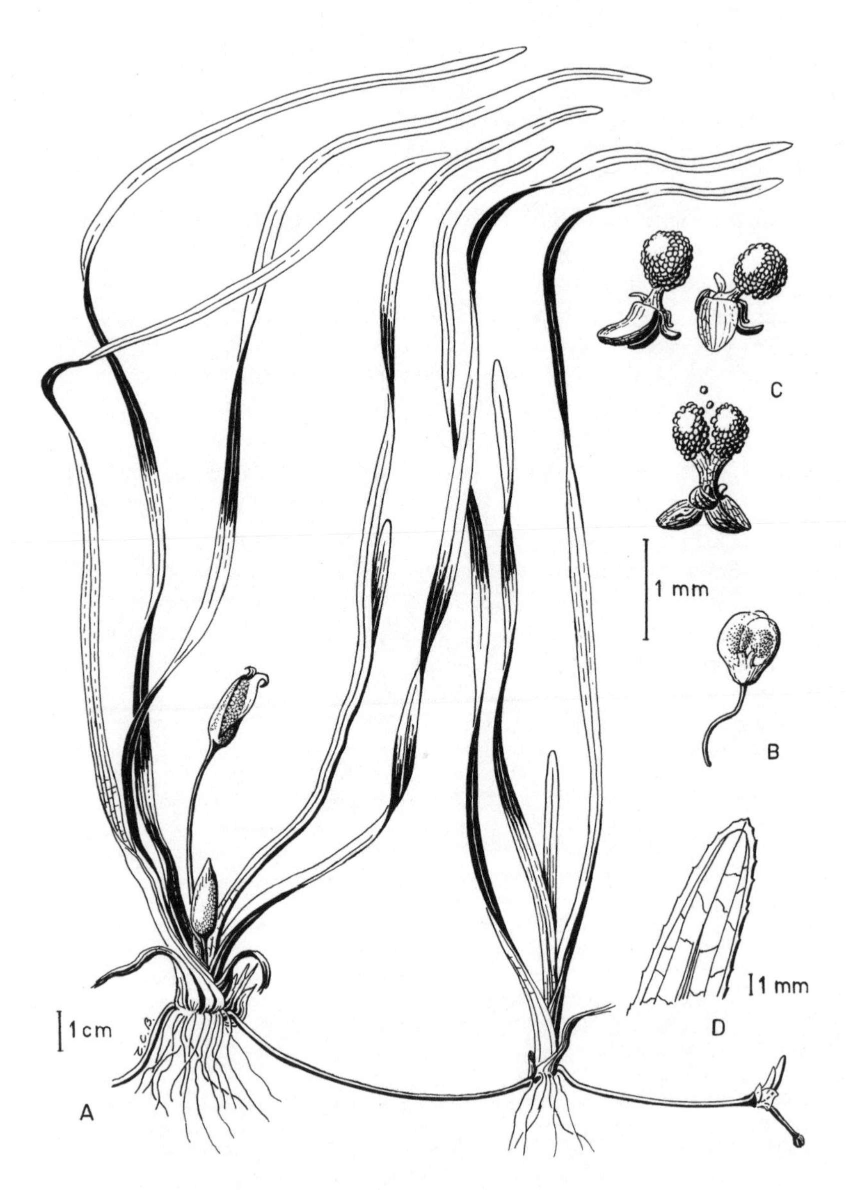

Figure 7. *Vallisneria americana* (staminate plant): A, shallow water plant in flower with closed and open spathes containing flower masses; B, staminate flower bud; C, open staminate flowers; D, leaf apex.

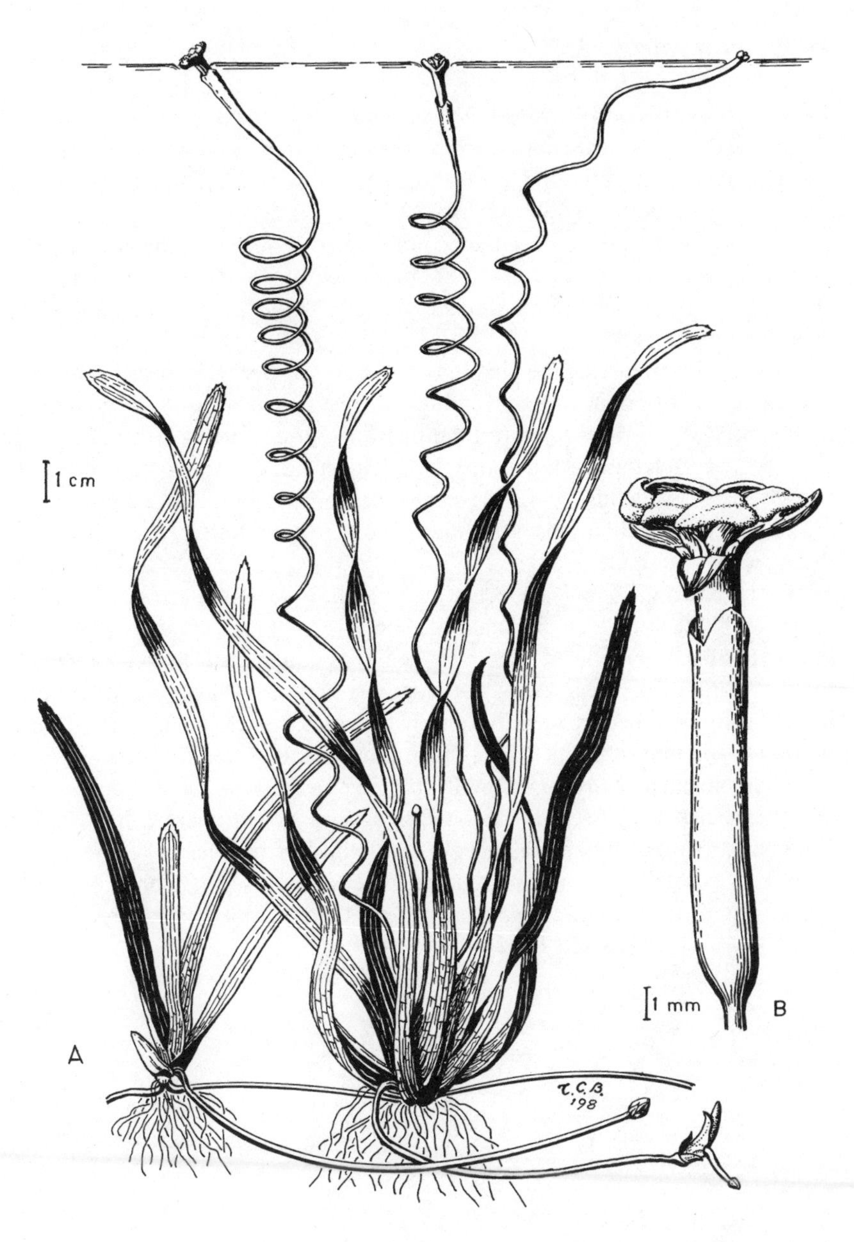

Figure 8. *Vallisneria* intermediate between *V. americana* and *V. spiralis* (pistillate plant): A, shallow water plant in flower; B, pistillate flower in spathe.

Vallisneria spiralis L. **European Tapegrass**

Vallisneria spiralis differs from *V. americana* as indicated in the key.

The pistillate plants are conspicuous when in flower; as the shiny, helically coiled peduncles that raise the flowers to the water surface, often in large numbers, are very visible from above. 2n = 20.

Staminate plants of *Vallisneria* are very seldom collected, perhaps because the inflorescence cannot be seen from above the water surface and the plants appear to be sterile. The few staminate plants examined from this region are clearly *V. americana*.

With pistillate material, the picture is not nearly so clear. The pistillate population of *Vallisneria* in British Columbia displays characteristics of both the above species in various mixtures, the incidence of any one decisive characteristic not correlating with that of others. The specific defining characteristics given by Svedelius (1932) do not work well for the population in this region, since no pistillate specimen examined here is completely typical of either species (see figure 8).

This situation encourages speculation that *Vallisneria* is not native in this province, but that the local population is the progeny of a number of separate incidents of introduction of a popular genus of aquarium plants. Following their liberation into local lakes, these species appear to have interbred and blurred their own specific identities in a hybrid swarm. Before one can reach a definitive solution to the problem, much more fertile material is needed for examination and perhaps for cultural experiment.

Both species of *Vallisneria* are found in eastern North America. In British Columbia, one or both species have been found in Christina Lake and in a number of coastal lakes, in water around a metre deep (see map on p. 53). In other parts of their ranges, *Vallisneria* species are regarded as an important food for waterfowl (Fassett 1957).

The Family Alismataceae (Alismaceae) — Water-plantain Family

Annuals or perennials with basal leaf rosettes and terminal cymose inflorescences of peculiar structure.

The earliest leaves produced are ribbonlike, without broad blades. Leaves produced later in the growing season usually have broad blades.

Branching in the inflorescence is basically alternate, with one bract, branch and prophyll at each node. However, along the main axis, and sometimes the main branches, the failure of two out of every three internodes to elongate during growth results in the formation of whorl-like clusters (false whorls) of bracts and umbel-like clusters (false umbels) of branches (Charlton 1973). Flowers bisexual or unisexual; sepals and petals differentiated, the latter often showy. Stamens six or more and, when more, originating in clusters opposite the petals. Carpels six to many, separate, originating in rough whorls or clusters, not spirally arranged; in fruit becoming achenes, or rarely, two-seeded follicles. Seeds with horseshoe-shaped embryos and no endosperm.

A worldwide family of 13 genera, related to the Butomaceae. Some members of this family are highly plastic in their response to vagaries in their environment. After the development of the juvenile ribbonlike leaves, the likelihood of later leaves producing broad blades varies from species to species; and may be modified by the environment. In some species, such as *Alisma gramineum* and *Sagittaria cuneata*, deep submergence in mid to late summer may induce retention of, or a return to, the juvenile foliage form, with suppression of the expanded blade (Arber 1920).

Key to Genera

1a. Flowers bisexual, small. Carpels in a single whorl. Terminal flower present. .... *Alisma*

1b. Flowers unisexual, often showy. Carpels very many in a head. Terminal flower absent. .... *Sagittaria*

The Genus *Alisma* L. — Water-plantain

Perennial herbs with rosettes of basal leaves and terminal panicles of flowers. Extension of vegetative growth from axillary buds produces leafy rosettes that in favourable conditions may terminate in secondary panicles the same year. Stolons or rhizomes as seen in *Sagittaria* are not produced.

Flowers occur in cymose panicles made up of false whorls, each comprising three bracts and a number of axillary scales. In the axil of each bract a lateral branch arises. Each branch bears at its base a bractlike, two-toothed prophyll in the angle between the branch and the main axis. Branching from the axils of the basal and more distal prophylls increases the number of flowers originating from the axil of each bract in the axial whorl. The panicle ends in a terminal flower.

Each flower consists of three sepals, three white or, occasionally, pink petals, six stamens, two opposite each petal, and 10 – 25 separate carpels in a whorl. The margins of each carpel are pressed together, but not united, at flowering time (Singh and Sattler 1972). The fruit is a whorl of achenes.

Key to Species

1a. Leaf blades elliptic, and cordate, rounded or broadly cuneate at base. Panicle much taller than leaves. Petals about twice as long as sepals. Anthers about twice as long as their thickness. ...*A. plantago-aquatica*

1b. Leaf blades lanceolate to linear, tapered to bases and apices. Anthers little or no longer than their thickness. 2

2a. Leaf blades lanceolate, 10 – 20 cm long by 2 – 3.5 cm wide. Panicle ample, of four to eight whorls, much taller than leaves. Petals about twice as long as sepals. Achene widest at the middle, with a straight stylar beak arising from the ventral margin about 1/3 of the margin's length from the top. ....*A. lanceolatum*

2b. Leaf blade narrow and ribbonlike, or with a small lanceolate blade up to 7 cm long. Panicle from shorter to a little longer than the longest leaf, erect or bent over, of two to five whorls. Petals shorter than to slightly longer than sepals. Achene widest near the apex, with a curved stylar beak arising from the top of the ventral margin. ....*A. gramineum*

Alisma gramineum Lejeune Ribbon-leaved Water-plantain

***A. geyeri* Torrey *ex* Nicolet**
= *A. gramineum* var. *angustissimum* (DC) Hendricks
***A. plantago* var. *angustissimum* (DC)**
= *A. gramineum* var. *angustissimum* (DC) Hendricks

Leaves typically narrow and ribbonlike, floating or submerged, flaccid and trailing; or with water levels fluctuating or declining in late summer, producing stiff, emergent leaves with small lanceolate to elliptic blades up to 5 (rarely 7) cm long.

Panicle from shorter than to a little longer than the longest leaves, with two to five whorls, erect, or the axis and branches curving over and downward.

Flowers small, 3 – 4 mm wide, the white or pink petals shorter than to barely longer than the sepals. Anthers roundish, not or scarcely longer than their thickness. Carpels with recurved styles.

Achenes widest near the apex, with two or more dorsal grooves, and a strongly curved stylar beak up to 0.5 mm long, arising at the top of the ventral margin of the achene. 2n = 14. Figures 9 and 10.

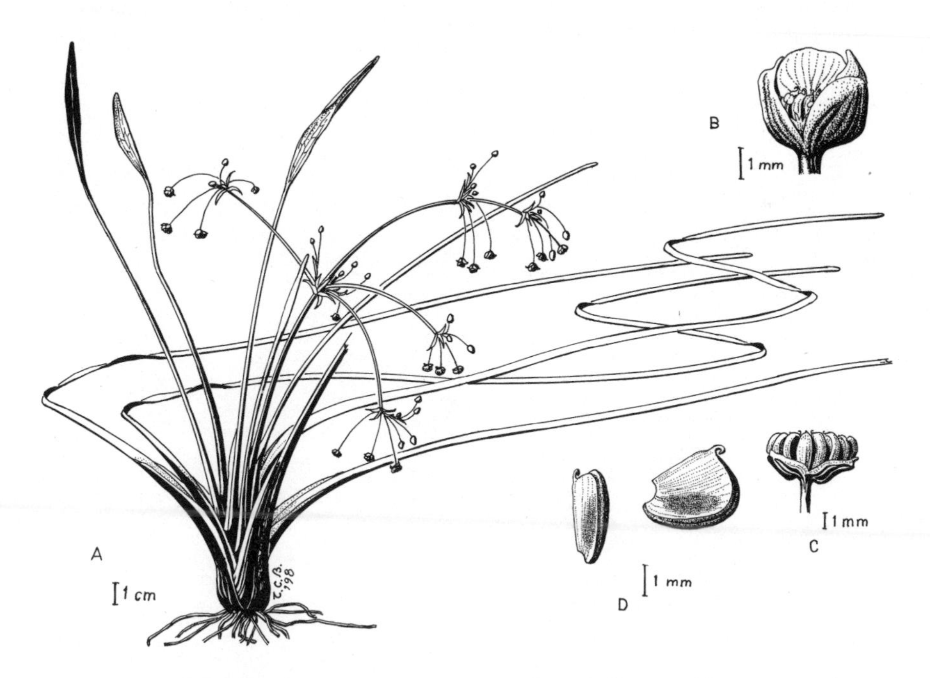

Figure 9. *Alisma gramineum* var. *gramineum*: A, plant; B, flower, with nearest petal partly removed; C, fruit (whorl of achenes); D, achenes.

Key to Varieties

1a. Leaves mainly submerged to floating, linear and ribbonlike, or with narrow lanceolate blades up to 7 mm wide. Panicle not overtopping the longest leaves, the axis and branches often curving over and downward...var. *gramineum*

1b. Leaves usually erect and emergent, with lanceolate blades up to 50 mm long by 5 – 20 mm wide. Panicle erect, usually overtopping the leaves...var. *angustissimum*

Our two varieties differ strikingly in appearance, but it is uncertain how much of the diversity is genetic in origin, and how much is due to the varying response of a very plastic species to varying depth of water or to cycles of flooding and emergence. Variety *angustissimum* appears to flower and fruit later than var. *gramineum.*

A plant of marshes, sometimes within the uppermost tidal range, or completely submerged in lakes or rivers, this species has been found flowering a metre beneath the surface of the upper Columbia River. Widespread in the northern hemisphere, it is seldom recorded in British Columbia. However, its scarcity may be more apparent than real, as it is easily overlooked in marsh vegetation, or when deeply submerged. Variety *gramineum* has been found in the southern interior of British Columbia, and var. *angustissimum* (DC) Hendricks in southwestern British Columbia from the Cascade Range to the coast, with one questionable record from Vancouver Island, and one from Shuswap Lake.

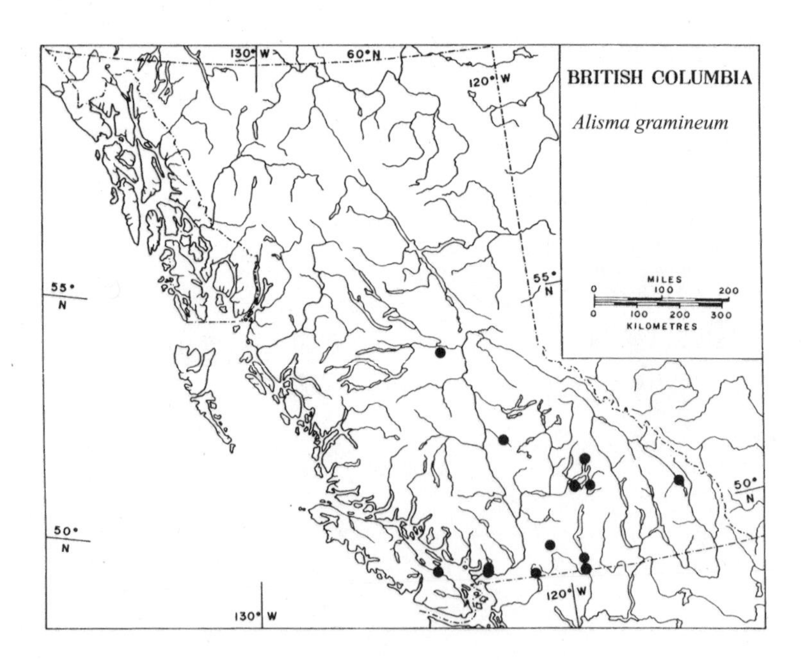

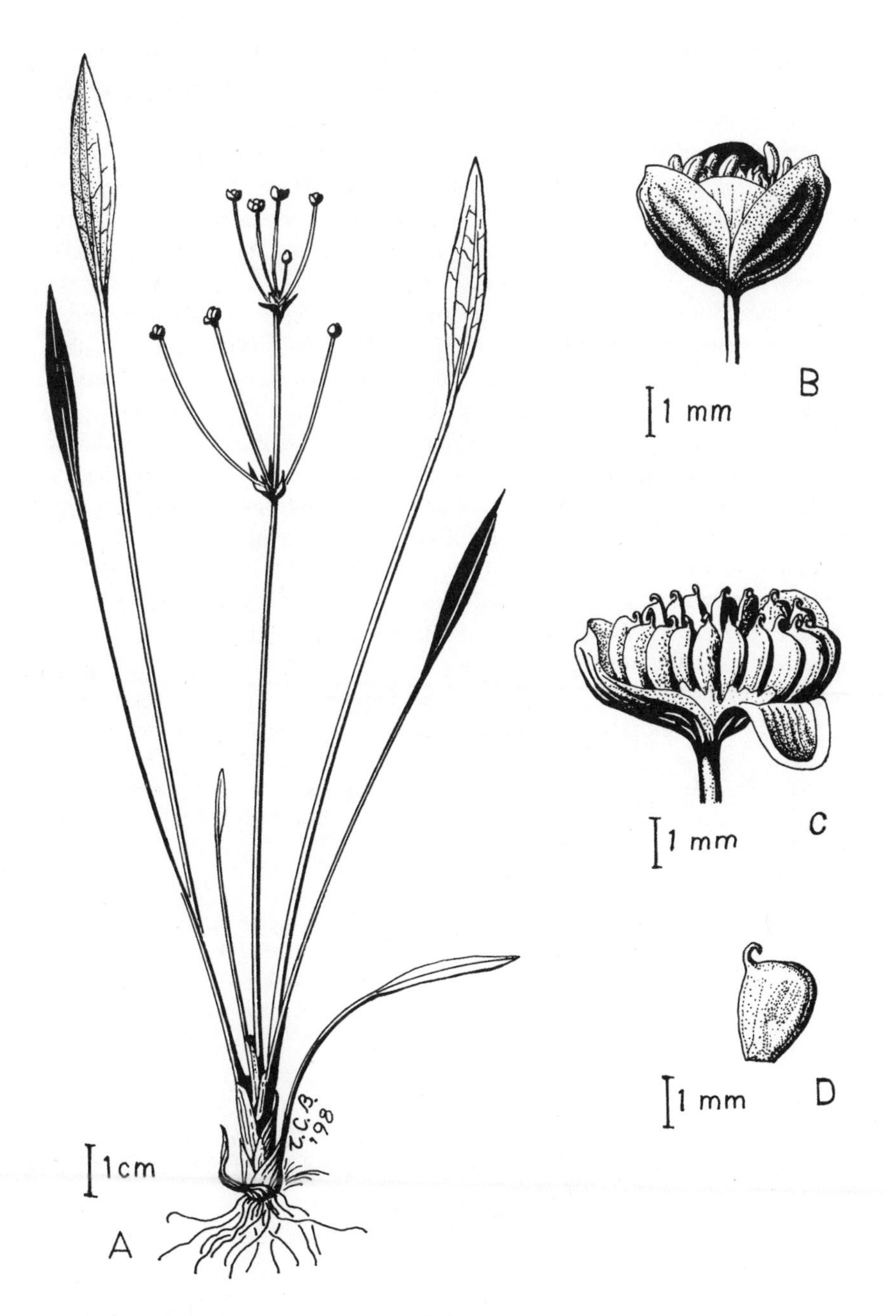

Figure 10. *Alisma gramineum* var. *angustissimum*:
A, plant; B, flower; C, fruit (whorl of achenes); D, achene.

Alisma lanceolatum Withering — Narrow-leafed Water-plantain

A robust perennial, up to at least 40 cm tall, similar to the much more common *A. plantago-aquatica*, but with lanceolate leaf blades 12 – 20 cm long by 2 – 3.5 cm wide, five to six times as long as their width; the inner lateral veins diverging from the midvein a fifth to a quarter of the length of the blade above its base.

Panicle with four to eight bracted whorls of branches, and standing much taller than the leaves. Flowers about 6 mm in diameter, with pink petals rather sinuately margined, rounded or pointed. Stamens normally six, the anthers a little longer than their width. Carpels 14 – 18 in a whorl.

Achenes flat, widest at or near the middle; with one or two dorsal grooves; the faces rather concave, and translucent so that the seed is discernible; the style straight or nearly so, about 0.5 mm long, arising from about two-thirds of the way up the ventral margin of the achene. $2n = 26$. Figure 11.

Native to Europe, and introduced widely in North America. In British Columbia, reported at Vancouver, Coquitlam and in the Fraser River delta, in ditches (material at the University of British Columbia); and at Enos Lake near Nanoose Bay on Vancouver Island (the last of the basis of incomplete material at the University of Victoria). In Europe this species is reported to form hybrids with *Alisma plantago-aquatica* (Dandy 1980, Stace 1997).

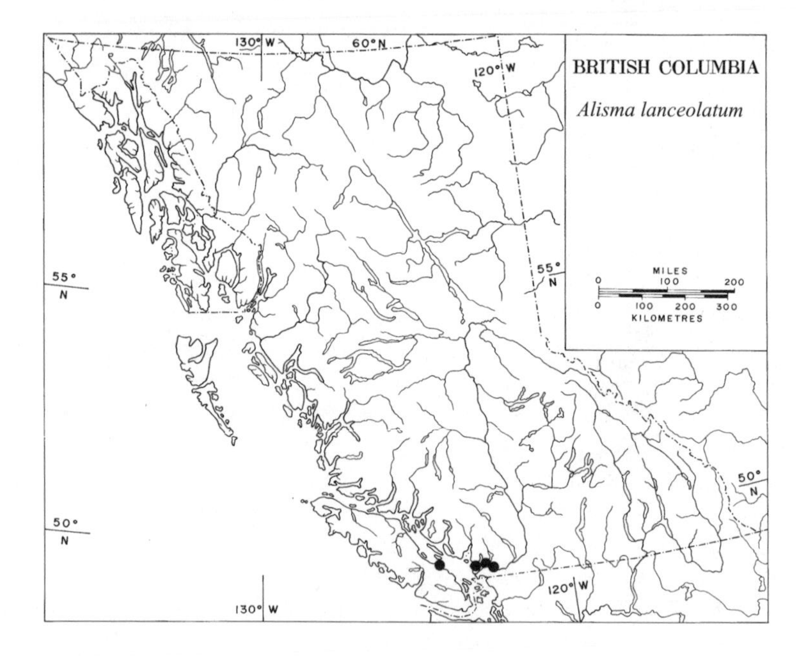

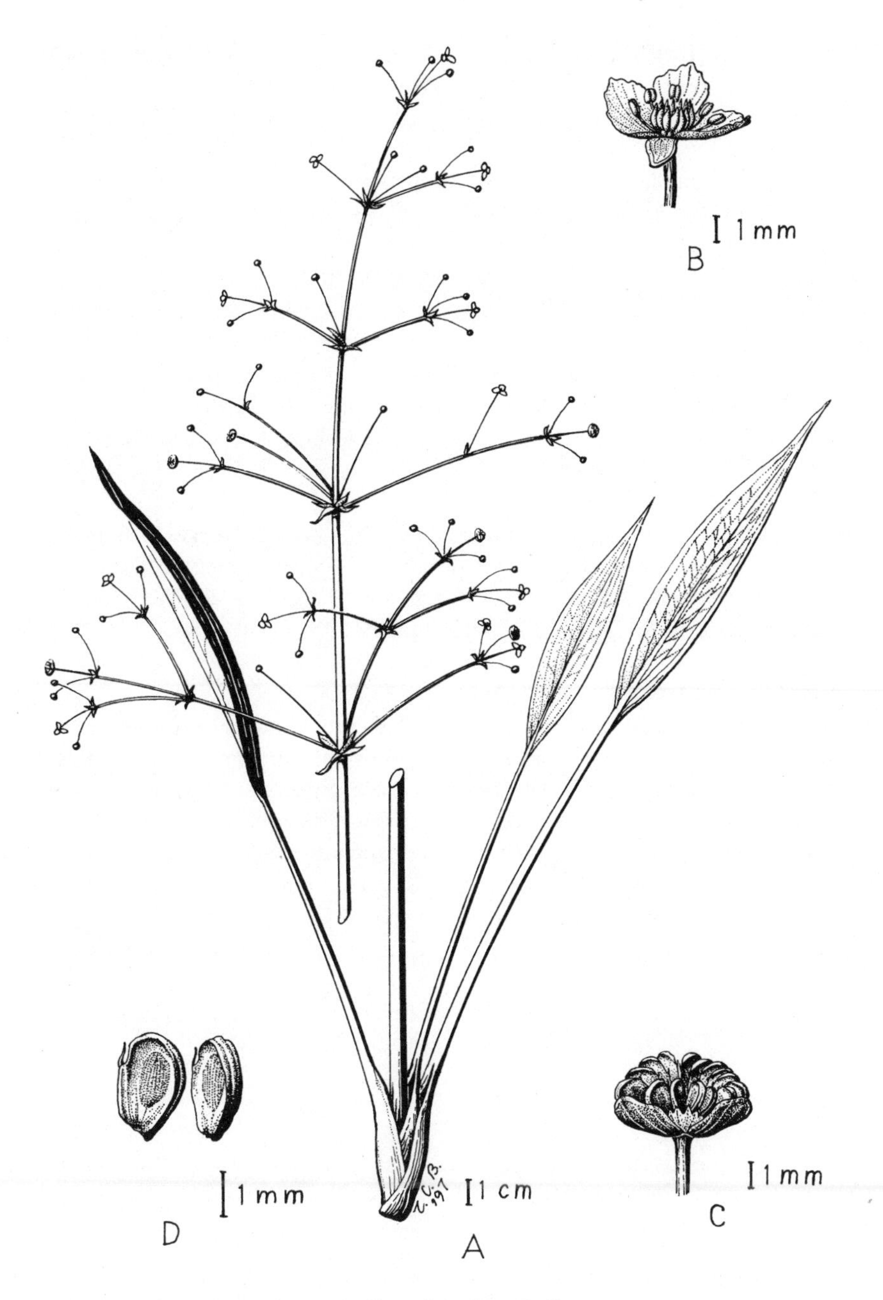

Figure 11. *Alisma lanceolatum*: A, flowering plant; B, flower; C, whorl of achenes; D, achenes.

Alisma plantago-aquatica L. **Water-plantain**

***A. triviale* Pursh = *A. plantago-aquatica* var. *americanum* Schultes and Schultes**

Erect plant, up to a metre tall when mature, with rosettes of upstanding leaves with elliptic, rarely lanceolate, blades 8 – 25 cm long by 3 – 15 cm wide; usually at least half as wide as its length; cordate, rounded or cuneate at base, the inner lateral veins diverging from the mid vein about 1/10 to 1/8 of the length of the blade above its base. Petioles semi-terete, channelled, and longer than the blades.

Panicles stiffly erect, much taller than the leaves, with four to eight false whorls of bracts and branches. Flowers 6 – 9 mm in diameter. Petals twice as long as the sepals, white or sometimes pink. Anthers twice as long as their thickness. Styles straight or curved, usually arising high on the inner margins of the carpels.

Achene 3 – 4 mm long, with one or two dorsal grooves; the faces plano-concave, opaque or translucent so that the seed is discernible; the stylar beak usually straight, slender and fragile, 0.5 to 0.7 mm long, arising between half and two-thirds of the way up the ventral margin. 2n = 14. Figure 12.

Key to Varieties

1a. Petals predominately white. Leaf blade elliptic, cordate to rounded at base. Style subterminally attached to ovary, curved at flowering time, curved to straight in fruit, 0.5 – 0.7 mm long. ....var. *americanum*

1b. Petals pink. Leaf blade relatively narrower than in var. *americana*, broadly lanceolate to narrowly ovate, cuneate to cordate at base. Style laterally attached to ovary, straight at flowering and fruiting times, attached midway along the ventral margin of the achene. ....var. *plantago-aquatica*

The styles in this species are very fragile, and are frequently broken off the achenes on dried fruiting specimens.

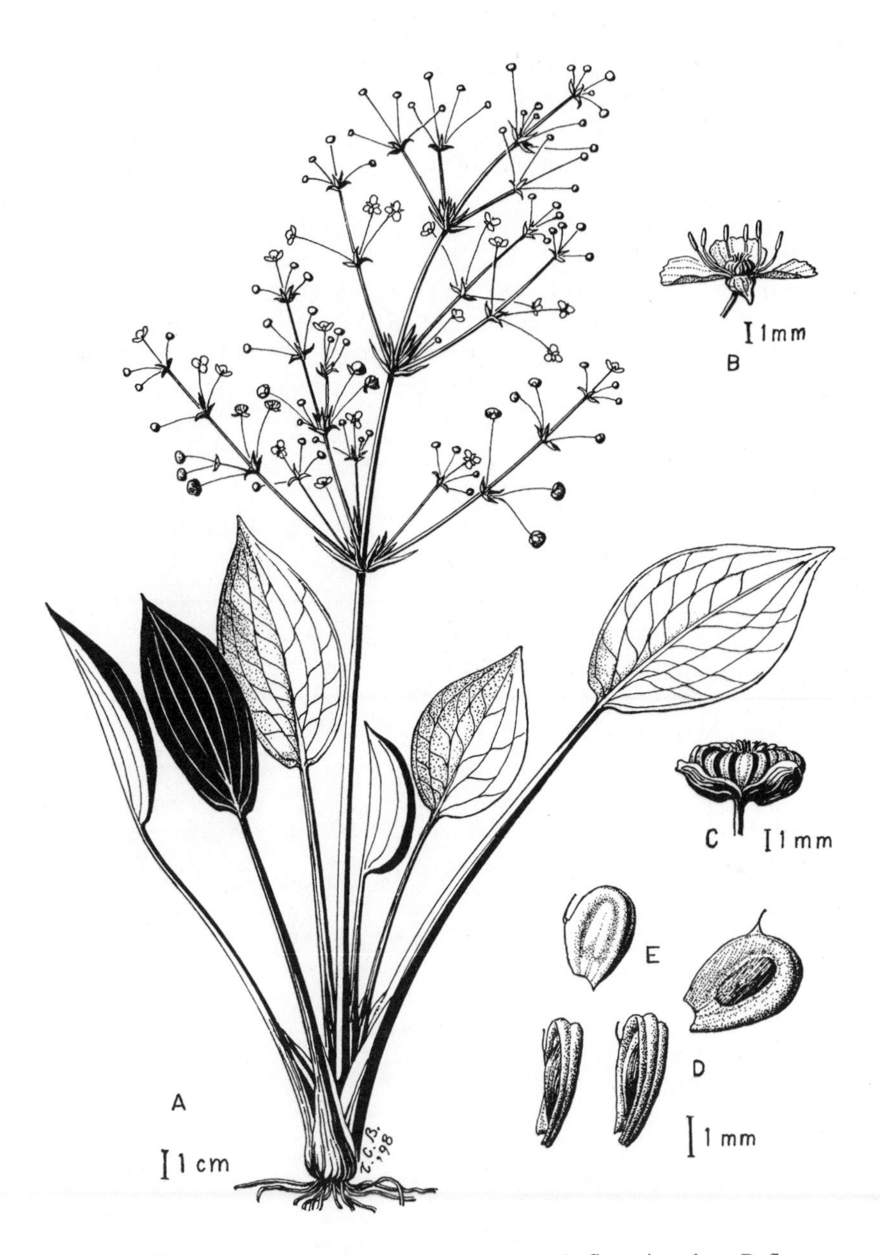

Figure 12. *Alisma plantago-aquatica* var. *americanum*: A, flowering plant; B, flower; C, whorl of achenes; D, three achenes; E, achene of var. *plantago-aquatica* (to same scale as D).

This species is of nearly worldwide distribution. Our common North American variety, var. *americanum*, is widespread in British Columbia, reaching the northern border of Alberta and extending eastward to Nova Scotia and southward to California. Usually a plant of muddy ditches and marshes, it is seldom found in deep water.

Variety *plantago-aquatica* is the typical variety in Eurasia, and has been recorded occasionally as an introduced plant at several points in North America, including Seattle, Washington. It has recently been found on Bowker Creek at Victoria, and may be found elsewhere in this province. It is desirable to collect leaves, flowers and fruits together, since the incidences of the distinguishing characters, as described in the key, are not perfectly correlated with each other.

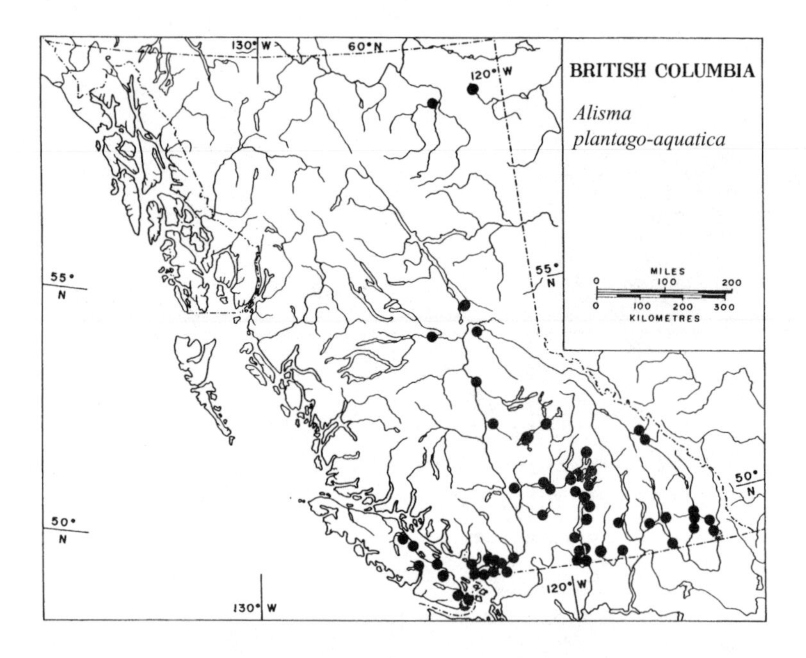

The Genus *Sagittaria* L. — Arrowhead

Perennial herbs with rosettes of leaves and terminal inflorescences. Extension of vegetative growth is from axillary buds that give rise to slender rhizomes or stolons that may terminate in overwintering starchy tubers.

The inflorescence is a raceme, or sometimes a panicle, with bracts and axillary pedicels in false whorls of three. Prophylls and their axillary branches are not normally produced, and lateral branches bearing more than one flower are seldom found. There is no terminal flower.

The flowers are of two kinds: staminate flowers occupying the upper part of the inflorescence, and pistillate flowers the lower part. Occasionally plants are found that are entirely staminate. Each flower has three sepals and three relatively large white petals. Staminate flowers produce many stamens; the outer ones in pairs opposite the petals, the inner in whorls alternating with the outer ones; and a number of abortive carpels. Pistillate flowers have several staminodes (sterile, abortive stamens) and many carpels in rough whorls, forming a globose head; each carpel with closed and united margins and with a short style and stigma (Singh and Sattler 1973). The fruit is a globose head of many flattish achenes, each with a short flat stylar beak.

Key to Species

1a. Living petioles usually angular in cross-section. Pedicels of equal length. Bracts ovate, rounded or hooded at tips, up to 10 mm long. Achenes with longish oblique beaks. .... *S. latifolia*

1b. Living petioles usually rounded in cross-section. Pistillate pedicels shorter than staminate ones. Bracts lanceolate, acuminate, up to 30 mm long. Achenes with short, apically pointing beaks. ....*S. cuneata*

Where abundant, members of this genus may be important wildlife food plants. The tubers are eaten by muskrats and other rodents (Fassett 1957) and, if not too big, by waterfowl. The upper plant parts are also eaten.

Sagittaria cuneata Sheldon — Arum-leafed Arrowhead

S. *arifolia* Nuttall

Perennial, spreading by slender rhizomes. Leaf form exceedingly plastic, varying in response to depth of water and other influences. Early leaves reduced to tapering flat petioles without blades; later leaves with usually rounded petioles when alive, floating or emergent and erect with sagittate blades with narrow or broad basal lobes, or occasionally ovate and lacking the basal lobes. In deep water the plant develops very long filiform petioles with very small floating ovate blades.

The raceme bears lanceolate acuminate bracts up to 30 mm long. Staminate pedicels are about twice as long as the pistillate. Petals are twice as long as sepals.

Achene 2.5 – 4 mm long, with an apical beak 0.2 – 0.5 mm long, projecting in line with the axis of the achene; pronounced dorsal and ventral keels that are almost winglike, and rather translucent walls over the locule, in which the embryo, closely invested by the seed coat, can be distinguished with the aid of a low-powered microscope. 2n = 22. Figure 13.

Distributed widely across North America, from central Alaska, Yukon and southwestern Northwest Territories (Porsild and Cody 1980) eastward to Nova Scotia, and southward to California and Texas. This is the commoner and more widespread of our Arrowheads in this province, where it is found all over the interior, but is almost absent from the coast. It occurs in ditches and in

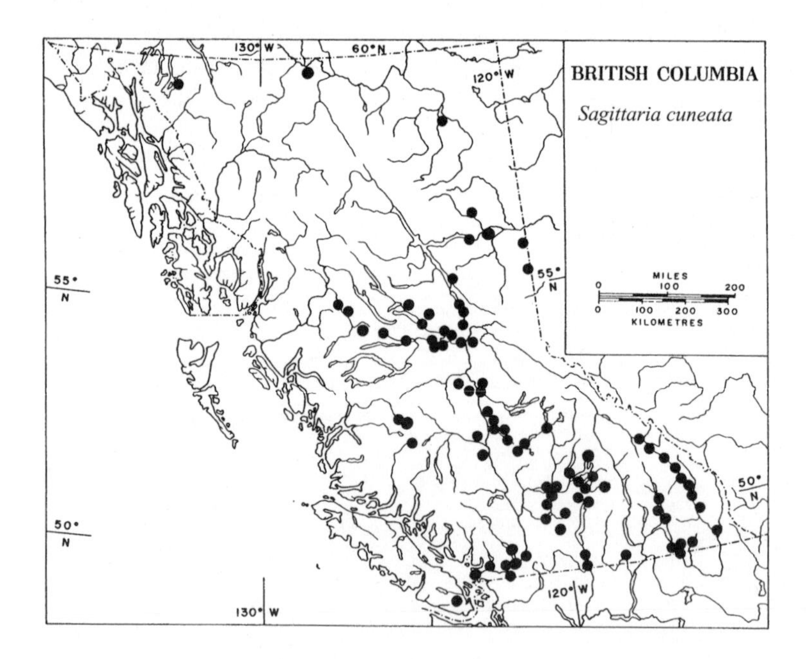

sheltered, or occasionally open water, to a metre or more deep, where the leaves and inflorescences may barely reach the surface. I have not seen it flower in water more than a metre deep.

Figure 13. *Sagittaria cuneata*: A, plant with staminate flowers open and pistillate flowers at fruiting stage; B, achene.

Sagittaria latifolia Willdenow — Broad-leafed Arrowhead, Wapato

Perennial, spreading by slender rhizomes often ending in tubers. Leaves apparently less plastic in form than those of *S. cuneata*; the mature plant normally with erect sagittate blades with lobes of varying width, on petioles that are angular in cross-section when alive.

Monoicous or occasionally dioicous: purely staminate plants are not uncommon in our population. The inflorescence appears more slender than that of *S. cuneata*, due to the shorter staminate pedicels, which are no longer than the pistillate ones. Peduncle triangular in cross-section. Bracts 5 – 10 mm long, rounded or hooded at tips. Petals twice as long as the sepals. Achene 3 – 5 mm long, with an oblique beak 0.5 – 1.5 mm long, turned toward the apex of the flower axis. 2n = 22. Figure 14.

Widespread in the Americas, from central British Columbia to James Bay and Nova Scotia, and southward to California and tropical regions. This species is uncommon in British Columbia, but is found at the coast as well as in the interior. Records from the Victoria area are of staminate plants, and must derive from recent introductions. Other Vancouver Island records may be similar in origin. This species is not seen in as deep water as is *S. cuneata*.

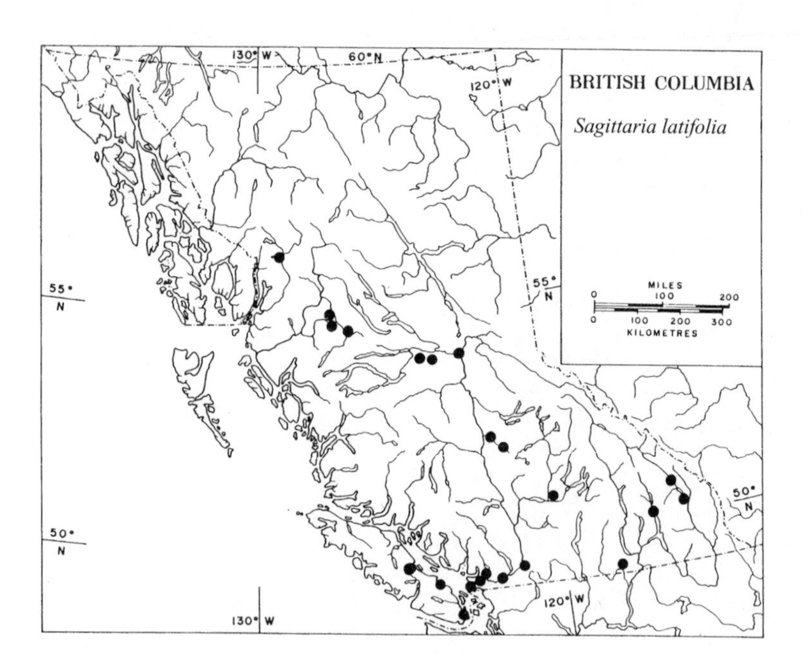

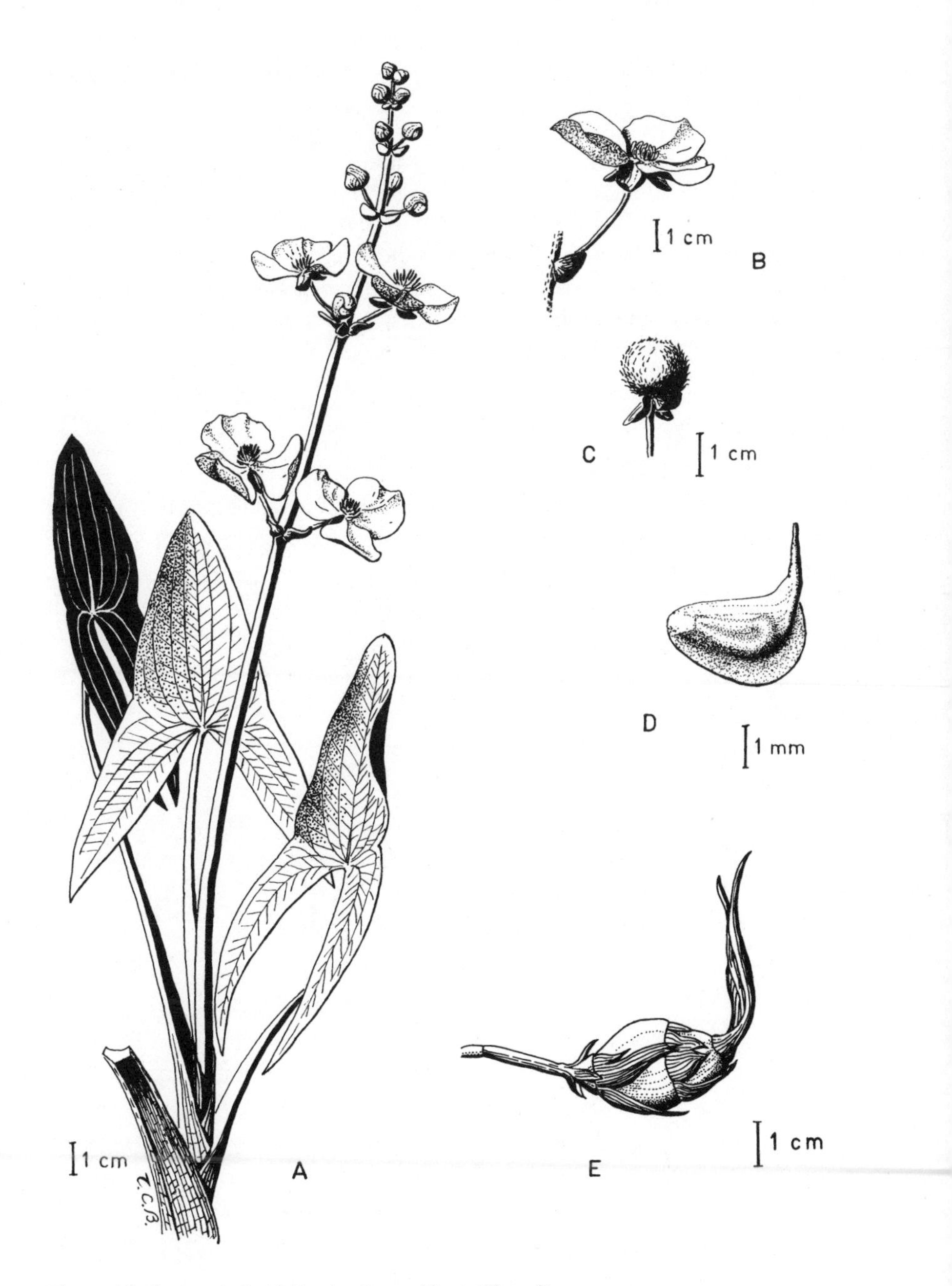

Figure 14. *Sagittaria latifolia*: A, plant with pistillate flowers open; B, staminate flower; C, head of achenes; D, achene; E, tuber.

The Family Scheuchzeriaceae — Scheuchzeria Family

Rushlike bog plants with short rhizomes bearing persistent leaf bases and leafy lower stems. Leaves with prominently ligulate sheathing bases, semi-cylindrical blades and apical pores, with axillary scales represented by hairs (Tomlinson 1982).

Inflorescence a terminal, bracted, few-flowered raceme of trimerous flowers. Each flower arises in the axil of a bract, and bears six tepals, six stamens, and three (rarely more) separate carpels, each containing one to three ovules.

The fruits are widely spreading, stubby-beaked, inflated follicles, each containing one to three seeds. Seeds straight, ellipsoidal, without endosperm.

The family contains one genus of one species: *Scheuchzeria palustris*, widespread in the colder parts of the northern hemisphere.

The genus *Scheuchzeria* is sometimes treated as a member of the Family Juncaginaceae, as by Fernald 1950. The treatment here, placing it in a family of its own, follows the example of Hitchcock et al. 1969.

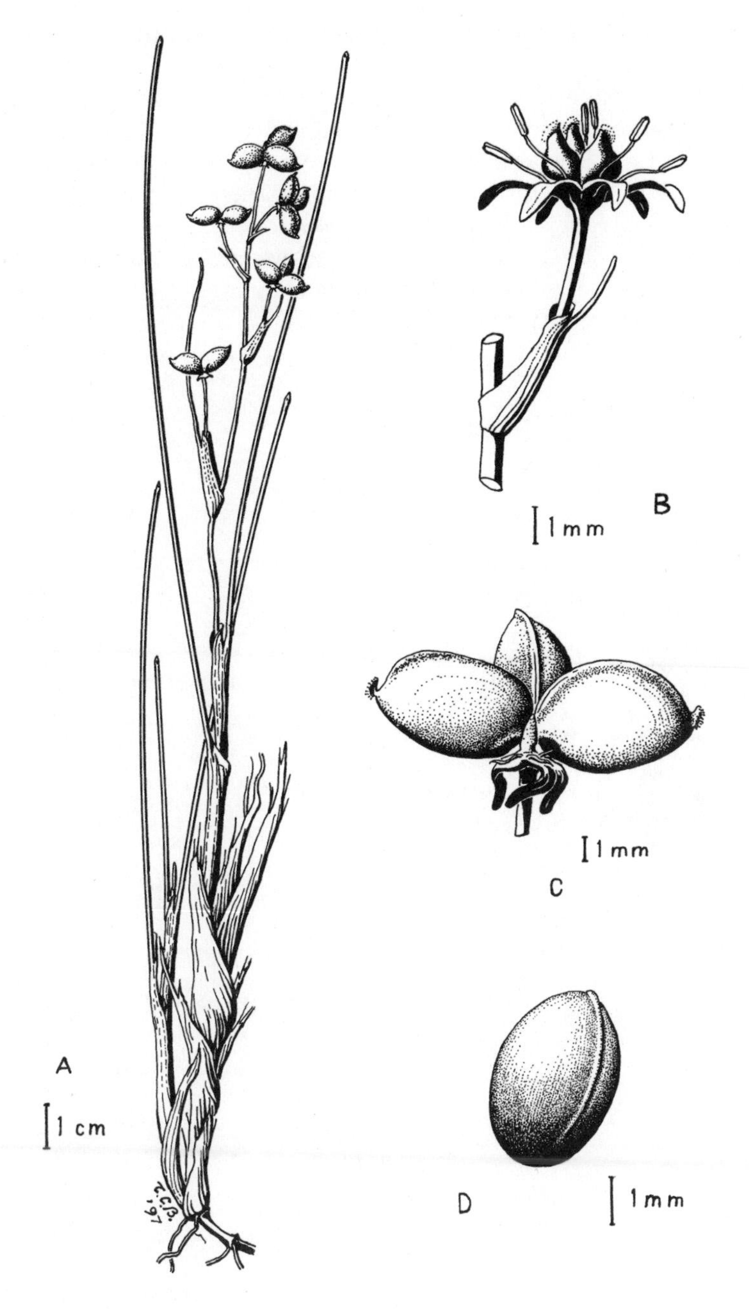

Figure 15. *Scheuchzeria palustris*: A, fruiting plant; B, flower and bract; C, fruit (three follicles); D, seed.

Scheuchzeria palustris L. var. *americana* Fernald

Growing 10 – 40 cm high, with pale greenish white flowers 5 – 6 mm across, with lanceolate, acute or obtuse tepals, and follicles 5 – 10 mm long with a beak up to 1 mm long; containing seeds 4 – 5 mm long. 2n = 22. Figure 15.

This species is of circumboreal distribution. The North American variety *americana* ranges from Alaska through the southwestern Northwest Territories to Newfoundland, and southward to New Jersey and California. The Old World var. *palustris* differs from our variety in having rather smaller flowers, fruits and seeds. It is found in pools in acidic *Sphagnum* bogs.

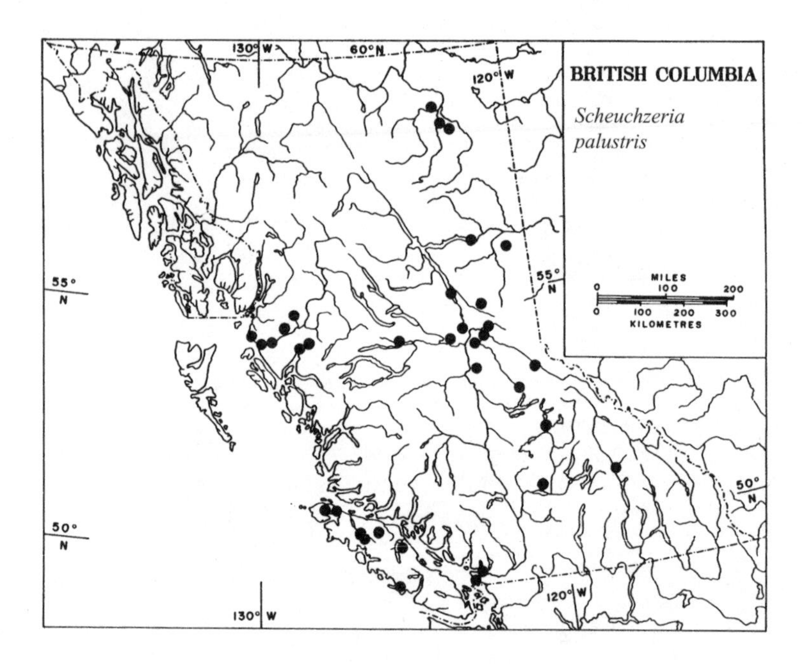

The Family Juncaginaceae — Arrow-grass Family

Rushlike marsh plants with erect basal rosettes of terete or rather flat linear leaves with sheathing bases.

Flowers mainly in spikes or racemes (some axillary), minute, with or without bracts or perianths. One to six stamens; one to six carpels, connate, with one ovule per carpel. Seeds with straight embryos and no endosperm.

Key to Genera

1a. Flowers in bractless racemes, all complete, with six perianth parts (tepals), and three or six carpels. Fruit a schizocarp............*Triglochin*

1b. Perfect and staminate flowers in minutely bracted spikes, with one carpel if any. Pistillate flowers in basal leaf sheathes, long-styled. Fruit an achene. ....*Lilaea.*

The Genus *Lilaea* Humboldt and Bonpland — Flowering Quillwort

Annual with terete leaves with ample basal sheaths, and flowers and pedunculate spikes arising in the leaf axils.

Flowers of three kinds: (a) perfect (bisexual) flowers in the lower spike, each flower with a minute bract (interpreted by some as a vestigial tepal), one stamen, and one carpel with a short style and a knoblike stigma; (b) pistillate flowers solitary in the basal leaf sheaths, and having apparently three united carpels and a very long style emerging from the leaf sheath and terminating in a knoblike and rather plumose stigma; and (c) staminate flowers toward the spike apex, with caducous bracts and subsessile two-chambered anthers.

The fruits are achenes with prominent apical stylar beaks, those from the basal flowers with three small projections near the tops of the ovaries.

Lilaea is sometimes treated as a family by itself, the Lilaeaceae, as by Douglas et al. 1969. The treatment here follows Hitchcock et al. 1969.

Lilaea scilloides (Poiret) Haumann — Flowering Quillwort
(*L. subulata* Humboldt and Bonpland)

Lilaea scilloides, the only species in this genus (figure 16), is found in shallow waters in alkaline, saline or brackish marshes. It is widespread in the Americas, but seldom collected. In North America it ranges from British Columbia, Alberta and Montana southward through the cordilleran region to Baja California. It also occurs in South America. In British Columbia, though not collected to date north of the Skeena River, it is probably more widespread than the current records indicate.

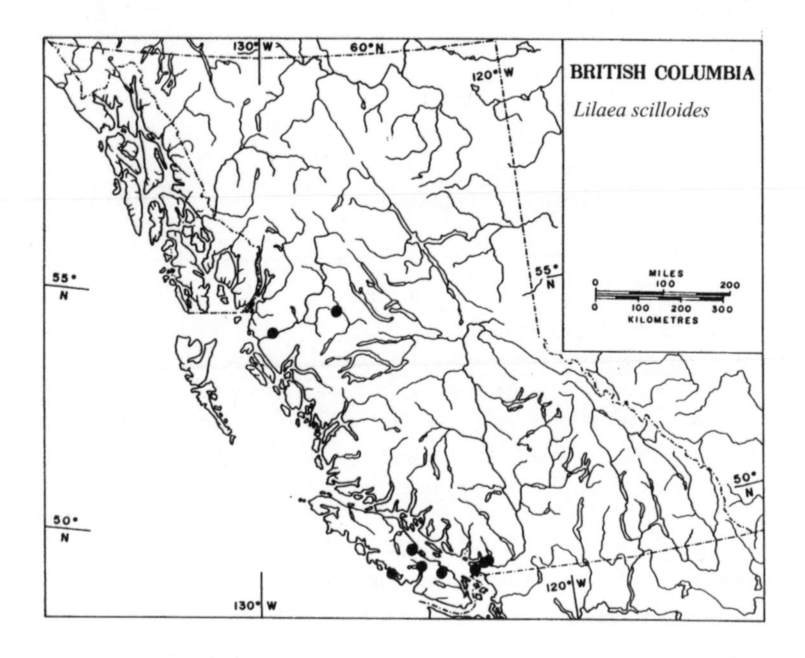

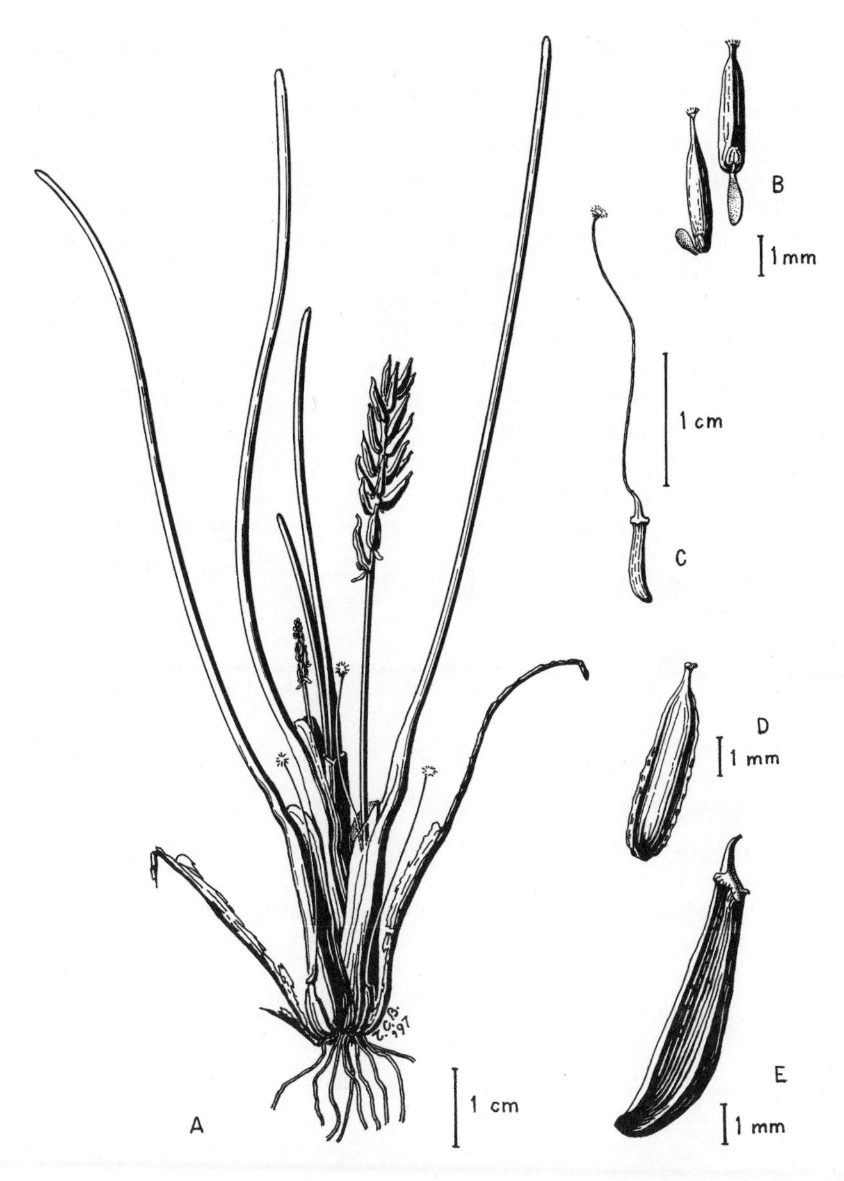

Figure 16. *Lilaea scilloides*: A, plant; B, perfect flowers, from lower part of spike; C, basal, axillary pistillate flower; D, fruit (achene) from spike; E, basal axillary fruit.

The Genus *Triglochin* L. — Arrow-grass

Marsh-inhabiting rhizomatous perennials with basal tufts of erect or arching linear leaves and scapose flowering stems.

Flowers all alike and bisexual, in terminal bractless slender racemes that sometimes appear lateral by reason of extension of vegetative growth from axillary buds. Each flower bears six similar tepals, six stamens, one opposite, and basally adherent to, each tepal, six connate carpels, of which three or all six may be fertile; each fertile carpel containing one erect ovule. Stigmas sessile or almost so, rather plumose.

Fruit a schizocarp; the fertile carpels at maturity splitting away from each other and from the central floral axis, and each containing one seed. Seed with straight embryo and no endosperm.

Species of this genus exhibit a tolerance for saline or alkaline environments. *Triglochin maritima* and *T. palustris* are well-known causes of livestock poisoning; the toxic effect being due to hydrocyanic acid (Kingsbury 1964).

The spelling of specific names in this genus is a subject of disagreement. The spelling adopted here follows that of Dandy 1980b.

Key to Species

1a. Fruit slender, tapering to base, the three fertile carpels splitting from a three-winged axis. Ligule divided.................................... *T. palustris*

1b. Fruit ovoid, rounded at base, the six fertile carpels splitting from a terete axis...2

2a. Ligule divided or clearly notched at apex, 0.5 – 1.5 mm long. Plant less than 20 cm tall. Raceme in our material usually not or scarcely taller than the leaves... *T. concinna*

2b. Ligule entire, 1.5 – 5 mm long. Plant 20 – 120 cm tall. Raceme distinctly taller than leaves... *T. maritima*

Triglochin concinna Davy **Graceful Arrow-grass**

(*T. maritimum* L. var. *debile* M.E. Jones = *T. concinna* Davy var. *debilis* (M.E. Jones) J.T. Howell, not *T. debile* (M.E. Jones) Löve and Löve)

A small slender plant, 5 – 30 cm tall, with slender interlacing rhizomes bearing few fibrous remains of old dead leaves.

Leaves slender, flat to semicircular in cross-section, blunt-tipped; the ligule 0.5 – 1.5 mm long, divided at its apex, forming two auricles.

Raceme relatively short and few-flowered in ours, often shorter than the leaves. Flower minute, about 1.5 mm wide and high at pollinating time, with six fertile carpels.

Fruit similar to that of the more common *T. maritima*, rounded at the base, the six fertile carpels splitting away lengthwise from the cylindrical floral axis and from each other. 2n = 48. Figure 17.

Key to Varieties

1a. Flowering shoot little if at all longer than the leaves, with 5 – 30 flowers. .... var. *concinna*

1b. Flowering shoot longer than the leaves, up to twice or more as long, with 20 – 75 or more flowers. .... var. *debilis*

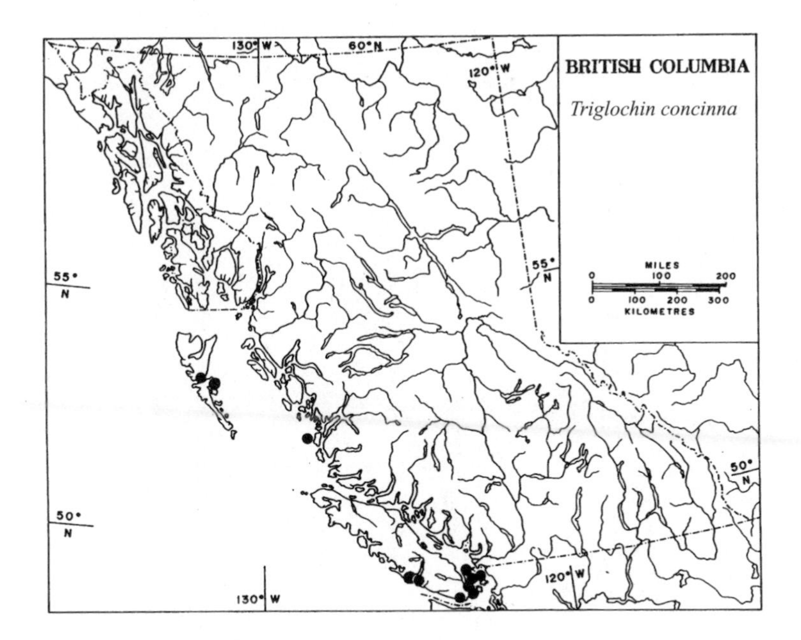

So far as is currently known, *Triglochin concinna* is represented in British Columbia only by its coastal variety *concinna*. It is found uncommonly, at scattered locations from the Queen Charlotte Islands to Baja California and in southern South America. It occurs in tidal marshes, forming, with its matted rhizomes, spreading patches of turf. Being a relatively inconspicuous plant,

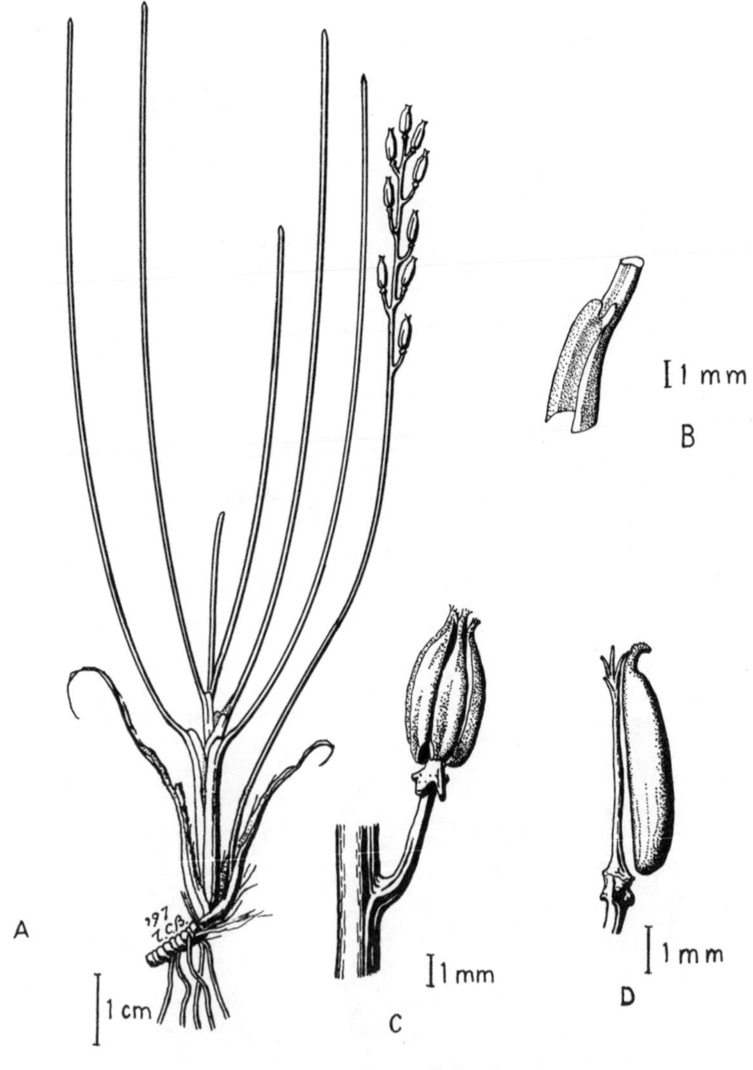

Figure 17. *Triglochin concinna*: A, fruiting plant; B, leaf sheath and divided ligule; C, fruit (six achenes): D, achene separating from floral axis.

this species can go unnoticed when growing in company with its taller relative, *T. maritima*; and may in reality be more abundant than existing collections suggest.

Variety *debilis* (M.E. Jones) J.T. Howell (1947) is an inland variety found in the Great Basin region of the western United States, but extending to Oregon, Montana, North Dakota and Colorado; the type specimen was collected at Johnson, Utah. This variety has not yet been found in British Columbia, but may be looked for in the southern interior of this province. The illustration (figure 18) is based on the type specimen of *T. maritimum* var. *debile*.

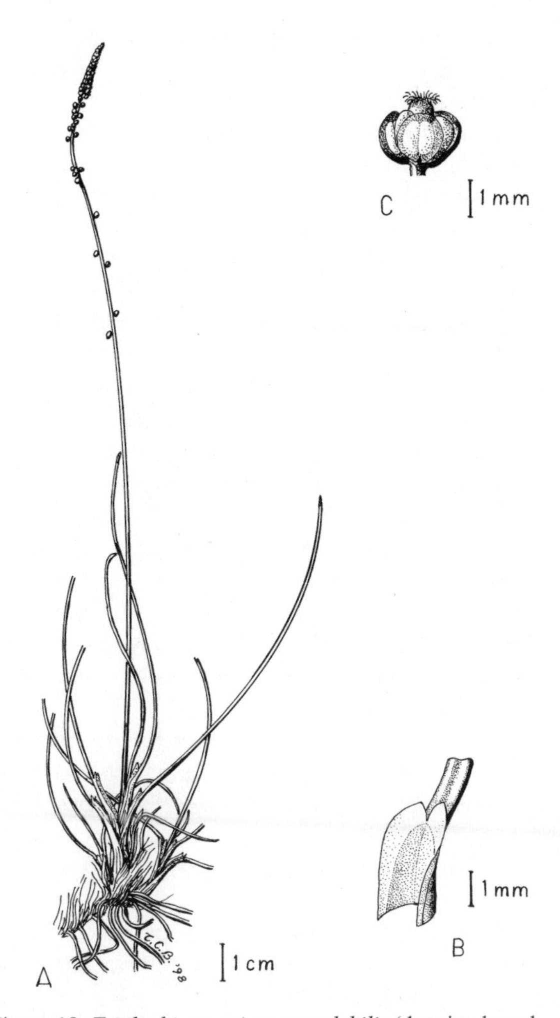

Figure 18. *Triglochin concinna* var. *debilis* (drawing based on the type specimen of *T. maritimum* var. *debile*):
A, flowering plant; B, leaf base and divided ligule; C, flower.

Triglochin maritima L. — Seaside Arrow-grass

(*T. debile* (M.E. Jones) Löve and Löve , based on *T. maritimum* var. *debile* M.E. Jones, and applied to the western Canadian population of *T. maritima*, is misapplied to this species.)

A taller and stouter plant than the other species of this genus; forming dense tussocks 20 – 120 cm tall, with a stout rhizome often covered with the fibrous remains of old leaf bases.

Leaf blades flat or sometimes semi-terete, 1.5 – 4 mm wide. Ligule (1.5 –) 2 – 4 mm long, entire, and rounded at apex.

Inflorescence a long raceme, usually well overtopping the foliage; the 50 or more flowers opening progressively from the base of the raceme toward the apex. Flowers 3 – 4 mm wide and high. Six fertile carpels (occasionally three), three of their stigmas often more prominent than the others.

Fruits rounded at base, ripening nearly simultaneously in each raceme; their constituent carpels splitting lengthwise apart and from the cylindrical floral axis. Figure 19.

Note on the Nomenclature

Triglochin maritima in the broad sense usually adopted by most botanists is treated by Löve and Löve 1958 as a complex aggregate species of circumboreal distribution as a whole, but with numerous regional micro-species or cytospecies, each distinguished by its characteristic chromosome number, but poorly distinguished by externally visible morphologic features.

The resulting classification, while correct in the strict biological and genetic sense, in that it defines the barriers to interbreeding, is of limited practical application in field botany and floristics. Considering the small number of published chromosome counts from western Canada, and in the interests of utility for the user of this volume, this author prefers to continue the use of the older classification based on visible characteristics. It is simply not practical to count the chromosomes of every specimen collected in order to identify it by name.

According to Löve and Löve 1958, all of our material previously called *T. maritima*, from Manitoba to the Pacific coast, has a diploid chromosome number (2n) of 96 and, therefore, should be called *T. debile* (M.E. Jones) Löve and Löve. Their statement was based on a very few widely scattered counts and may not reflect the full genetic complexity of the western North American population.

The name *debile* is misapplied in this species, since it properly applies to a variety of *T. concinna*. Examination of the type specimen of *T. maritimum* var. *debile* M.E. Jones reveals the very short, divided ligules that are characteristic of *T. concinna*, a feature not mentioned by Jones (1895) or Howell (1947), though the latter evidently saw it.

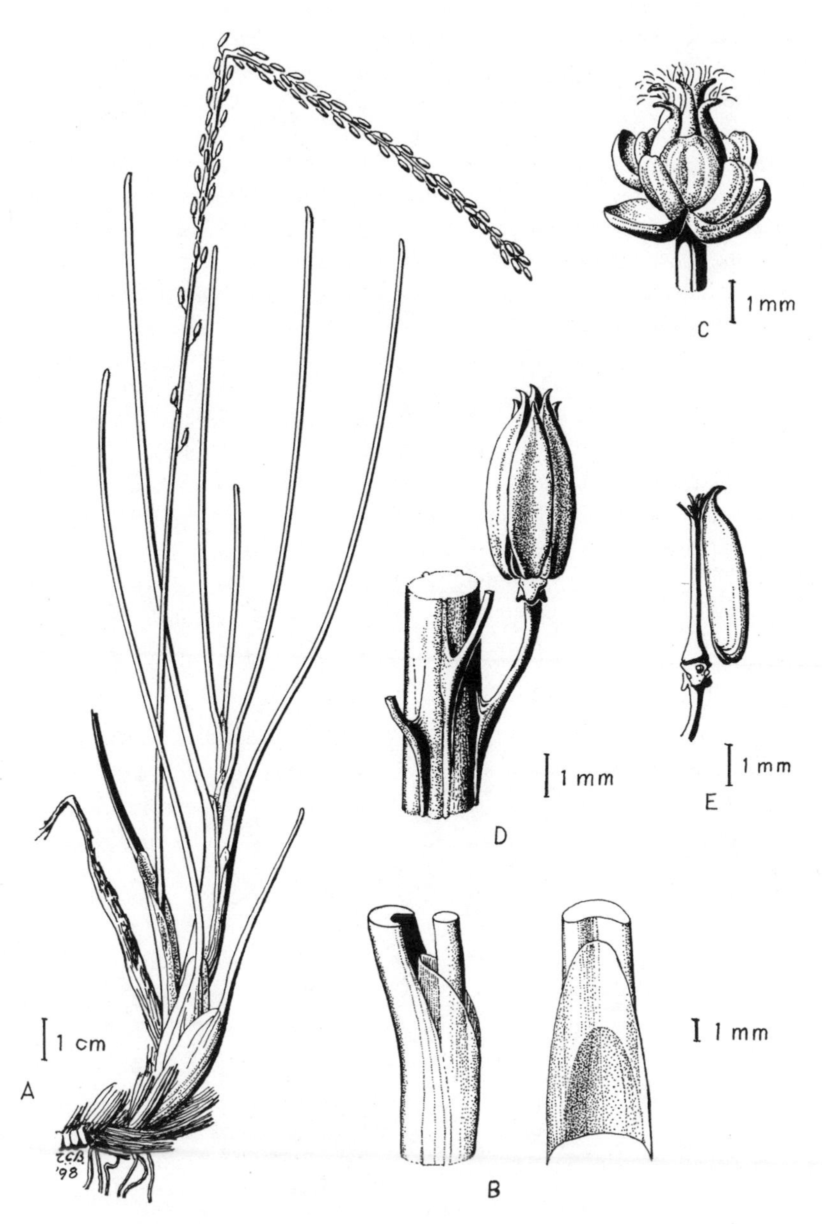

Figure 19. *Triglochin maritima*: A, small fruiting plant; B. leaf base and undivided ligule: two views; C, flower; D, fruit: a schizocarp of six nutlets; E, a nutlet separating from the floral axis.

A circumboreal species complex, *Triglochin maritima* is widespread in the Americas and Eurasia. Diploid (2n) chromosome numbers reported for the whole complex are: 24, 30, 36, 48, 96 and 144. British Columbian specimens from Tofino, on Vancouver Island (Pojar 1973) and the Yakoun River on the Queen Charlotte Islands (Taylor and Mulligan 1968) have shown 2n = 96.

T. maritima is the most common species of *Triglochin* in British Columbia. It is common in tidal marshes along the coast, and around brackish or alkaline ponds and seeps in the interior, forming large clumps.

A dwarfed, slender, very short-leafed plant, found at Osoyoos and Shuswap Lakes in the southern interior of British Columbia by A. and O. Ceska, superficially resembles the type specimen of *T. concinna* var. *debilis*. But on closer examination, its undivided, collarlike ligules show it to be a form of *T. maritima.* The illustration in figure 20 is based on a collection at the Royal British Columbia Museum by A. and O. Ceska from Osoyoos Lake, identified by them as *T. debilis* (collection no. V 170988: specimen nos 1 and 2).

If this plant is found to be a variety of genetic determination and thus worthy of recognition as a variety of *T. maritima,* and not just a response to environmental restraints, it will require a new varietal name. I am proposing no such name here, because more research needs to be done on this plant.

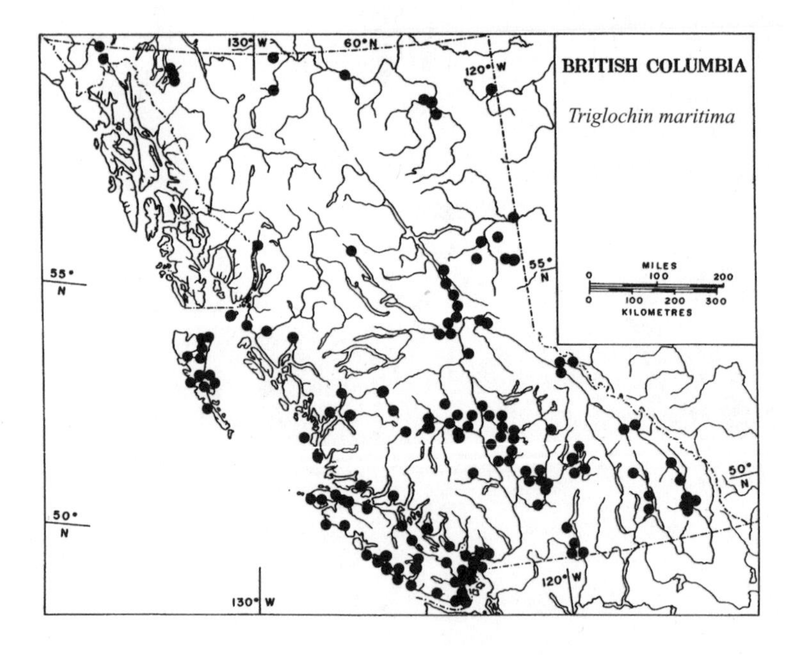

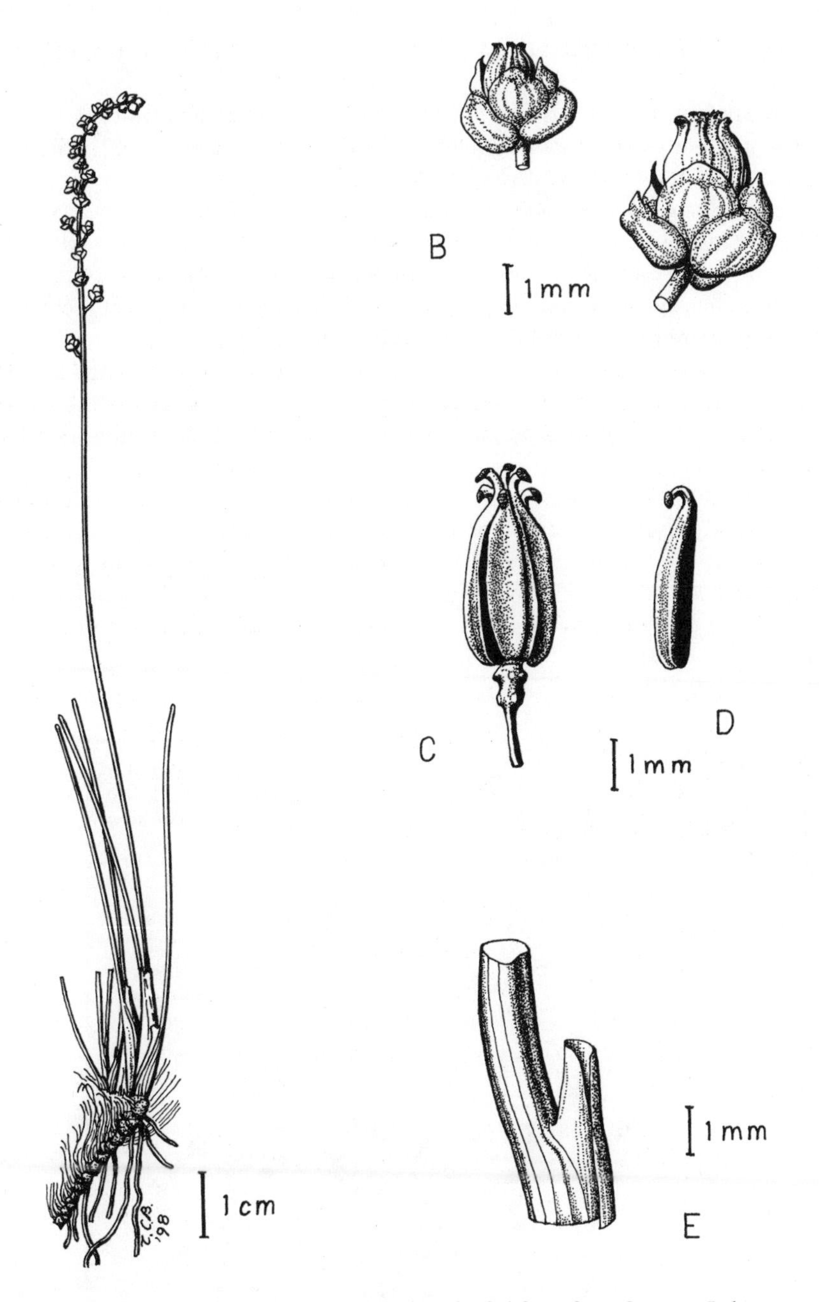

Figure 20. *Triglochin maritima* (dwarfed, short-leafed form from Osoyoos Lake; cỏllection V170988, specimens 1 and 2): A, plant; B, flowers: C, fruit (six achenes); D, achene; E, leaf base and ligule.

Triglochin palustris L. **Marsh Arrow-grass**

Slender plant, 15 – 60 cm tall, with short, weak rhizomes, and filiform, terete or semi-terete leaves, flat or channelled on the ventral or upper side. The ligule is 0.5 – 1.5 mm long, and divided at the apex.

Inflorescence is an apparently axillary, very slender, open, bractless raceme, the peduncle often flat when young. Flower minute, about 2 mm high and wide at pollinating time, consisting of 6 tepals, with a stamen opposite, and basally attached to, each tepal; the three fertile and three sterile carpels coherent into an ovary crowned by the sessile, brushlike stigmas.

The fruit is a slender schizocarp, tapering to its base; the three one-seeded fertile carpels separating progressively upward from their bases from the floral axis, which is flanged by the three adherent sterile carpels. 2n = 24, 28. Figure 21.

This species is widespread in Eurasia and the Americas, including Greenland, in saline, brackish, or alkaline marshes or seasonally moist meadows. In British Columbia, it is widespread across the interior, mainly in areas of restricted drainage in semi-arid regions, especially where the soil water may be brackish or alkaline. It is relatively uncommon on the coast in tidal marshes.

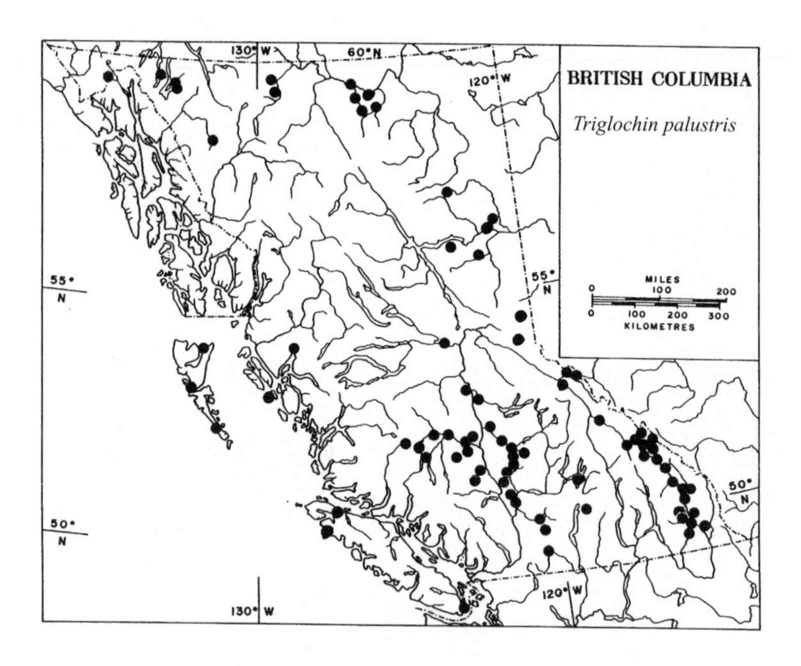

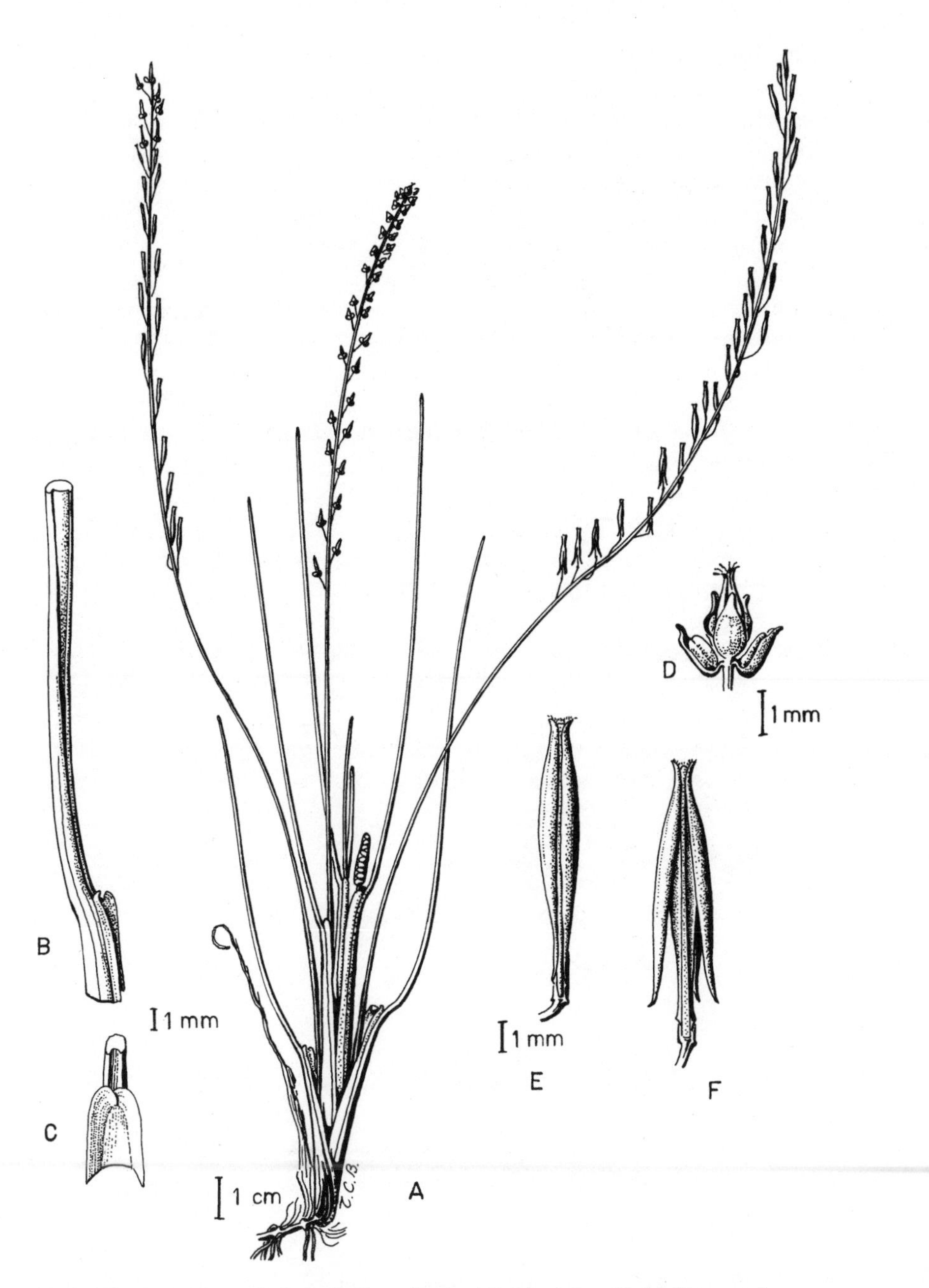

Figure 21. *Triglochin palustris*: A, flowering and fruiting plant; B, leaf base and divided ligule (side view); C, leaf base and divided ligule (face view); D, flower; E, fruit cluster (three achenes) from one flower; F, achenes separating from floral axis.

The Family Potamogetonaceae — Pondweed Family

Perennials, usually with slender submerged rhizomes or stolons, slender pliable stems that are submerged except for the inflorescences, alternate or sometimes subopposite leaves, and conspicuous ligular sheaths or ligules in our genera.

The inflorescence is a terminal spike with two to many alternate to apparently whorled, sessile, perfect flowers. Each flower has two or four stamens consisting of essentially sessile anthers, which are adnate to the bases of four conspicuous tepals in *Potamogeton*, while in *Ruppia* the two tepals are vestigial. There may be one to six (usually four) separate carpels, alternating with the stamens when of the same number; each carpel with a terminal stigma and containing one ovule. Each carpel matures into an achene, sometimes rather drupelike by reason of a spongy or fleshy outer layer. The stigma is sessile or on a stylar beak. The seed contains a curved or coiled embryo and no endosperm (Singh 1965).

The family, as treated now, contains five genera, two of which are represented in British Columbia.

Key to Genera

1a. Spike with two flowers. Stamens and tepals two per flower. Achenes with long stipes. ....*Ruppia*

1b. Spike with four to many flowers. Stamens and tepals four per flower. Achenes more or less sessile. ....*Potamogeton*

The Genus *Potamogeton* L. — Pondweeds

Soft-stemmed aquatic perennials of diverse aspect, totally submerged or with the inflorescence emergent and the uppermost leaves floating. Leaves alternate and two-ranked, or appearing subopposite under an inflorescence, as the distal stem internode fails to elongate. Totally submersed plants generally have leaves of uniform type, thin, usually sessile and often translucent. In species with floating as well as submersed foliage, the leaves differ strikingly in character. The floating leaves generally have longer petioles and broader, thicker, opaque blades with more longitudinal veins than have the submerged leaves. Transitional leaf forms sometimes occur. The sheaths and ligules are usually conspicuous and their varied characters are of diagnostic significance.

The flower spike bears four to many flowers, alternate or in irregular whorls or pairs. Each flower has four tepals, to each of which is adnate a sessile anther of two anther-sacs, and usually four separate sessile or subsessile carpels with terminal buttonlike stigmas. Pollen is transported by wind in most of our species; but in species of the subgenus *Coleogeton* (genus *Stuckenia*), it is transported by water on or beneath the surface (Dandy 1980, p. 10).

The fruit is a more or less sessile, often fleshy-walled achene, containing a U-shaped or coiled embryo, whose shape is often distinguishable through the distension of the outer layers of the fruit. The stigma may be sessile or terminate a distinct stylar beak.

Most species produce, in the bottom sediments beneath the water, slender horizontal rhizomes, which form the principal means of vegetative propagation. The rhizome, which bears reduced, scale-like leaves at its nodes, is formed from the first two internodes of a basal branch, which then turns upward to form an upright shoot with normal foliage. Buds in the axils of scale-leaves or of normal leaves at the base of the erect shoot may give rise to horizontal rhizome extensions in the same way (Tomlinson 1982). Distinct overwintering tubers are formed on the rhizomes of some species; a typical tuber consisting of two successive short, swollen internodes.

Many species produce special winter buds generally on the upper branches. These are very short branchlets with reduced leaves, and are easily detached. After the plant has decayed in the autumn, the winter buds lie on the bed of the water body over winter, sprouting in spring to produce new erect shoots. A few species, such as *P. crispus* and *P. foliosus*, produce both rhizomes and winter buds.

Potamogeton is a genus of about 100 species, of cosmopolitan range, but principally of the north temperate latitudes. About 40 species are indigenous to North America, 21 of them in British Columbia. The plants generally occur in still or slowly flowing fresh water; but a few species, such as *P. filiformis* and *P. pectinatus* are sometimes found in brackish lakes or saline waters of tidal marshes.

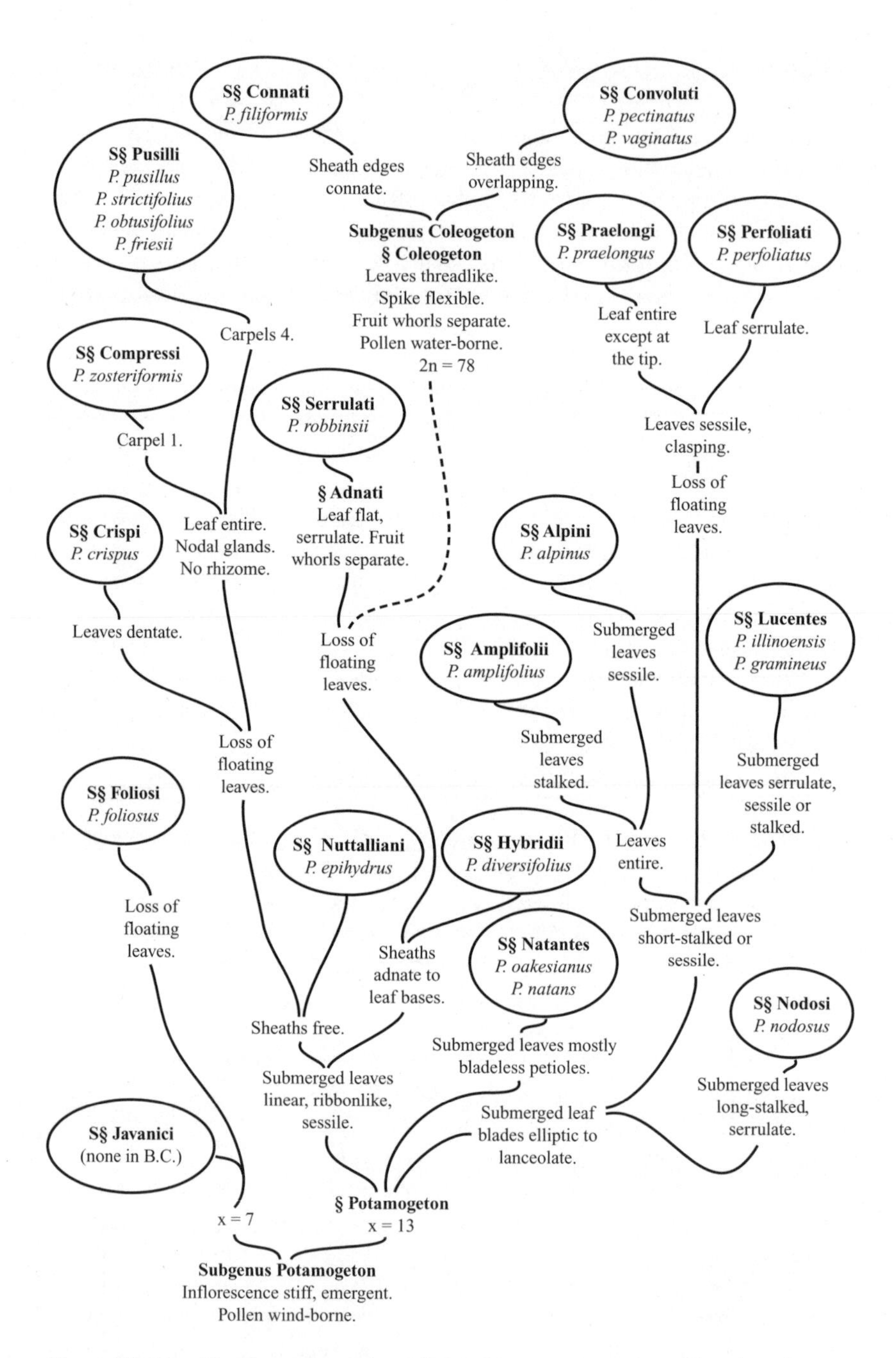

Figure 22. Classification and putative relationships among sections (§), subsections (S§) and species of *Potamogeton* in British Columbia. Modified and diagrammed after information in Les (1983).

Classification of *Potamogeton*

Systems of classification used in the past, as in Fernald 1950, have been based more exclusively on externally visible features than is the common practice today. Fernald (1950) treated *Potamogeton* as a member of the Family Zosteraceae, along with *Ruppia* and *Zannichellia*. Recent use of several new techniques has led to changes in the overall classification of this group. Analysis by Les et al. (1995) has led them to recommend the inclusion of *Zannichellia* in the Potamogetonaceae and the removal of *Ruppia* from that family. The striking resemblance between *Ruppia* and *Potamogeton* subgenus *Coleogeton* is thus seen as an example of parallel morphologic evolution in two genetically distinct groups in adaptation to the same environmental requirements. The classification of this whole group remains in a fluid state.

Within the genus *Potamogeton*, recent examination of the chromosome complements has revealed that there is a division of the genus into two series of species, having different basic chromosome numbers (denoted by "x") (Les 1983). This discovery has shown that some modifications of the existing system are needed.

One modifications pointed out by Les (1983) reflects the parallel trends toward the discard of floating forms of leaves with progressive adaptations to an immersed life in a number of separate evolving lines.

Another modification is the removal from the subsection *Pusilli* of *Potamogeton. foliosus*, in spite of their gross similarity, since *P. foliosus* belongs in the series with x = 7, while the remainder of the species in subsection *Pusilli* belong in the series with x = 13. *Potamogeton foliosus* differs from the species in subsection *Pusilli* not only in its chromosome complement, but in such morphological features as the possession of rhizomes, usually glandless nodes, and achenes with prominent beaks and serrated keels – features that are easy to overlook.

The scheme in figure 22 is a diagrammatic representation of the postulated relationships among the sections, subsections and species *Potamogeton* found in British Columbia. This scheme is based primarily on older works using visible morphological features as the main criteria for grouping species, but with modifications reflecting the newer information from Les 1983. In this scheme, the subsection *Foliosi* is proposed, to contain *P. foliosus*.

This scheme is tentative and no doubt temporary; and will in time be superceded by schemes better reflecting the genetic foundation of this genus, as more knowledge is realized through the increasing use of chemical analyses of genetic and other compounds.

The subgenus *Coleogeton* is sometimes treated as a genus, *Stuckenia* Borner (Holub 1997) or *Coleogeton* Les and Hayes (1996), separate from *Potamogeton*. Of these generic names, *Stuckenia* Borner has priority, having been published in 1912.

The separation is justified from a biological and genetic standpoint, based on these distinguishing features: the anatomy of its long, flexible peduncle; the lax submerged or floating inflorescence; water-borne rather than wind-borne pollen; the hexaploid chromosome complement (i.e., 2n = 78); and the absence of hybrids between its species and any other *Potamogeton* species, though hybrids do occur among the species in subgenus *Coleogeton*.

While I acknowledge the above distinctions in the interest of the utility of this book, I retain the subgenus *Coleogeton* here within *Potamogeton*.

The nomenclature of species in *Potamogeton* is in a most unsatisfactory state. Several of the species and species-complexes have circumboreal ranges, and were first described on the basis of European material, in European literature, and the type specimens, on which the original descriptions were based, are mostly in various European herbaria. Further, since the 19th century, the nomenclature applied to species and varieties of this genus has, to a great degree, evolved independently in Europe and North America. The resulting present confusion of names can only be sorted out properly by someone with access to the older European collections and libraries as well as those in North America.

The nomenclature used in this work is in general conformity with that accepted by most North American botanists as applicable to North American material. But its precision and correctness are questionable for several species and varieties, including *Potamogeton filiformis*, *P. friesii*, *P. gramineus* and *P. zosteriformis*.

Hybrids in *Potamogeton*

Hybrids have been recorded between many species in this genus. While of vigorous growth, they are usually sterile. However, their ability to propagate vegetatively from year to year can enable them to establish colonies of long duration and appreciable local extent.

These hybrids, as reported, are putative only; that is, they are founded on the examination and interpretation of character combinations in naturally occurring plants. I am unaware of any attempts to confirm the interpretations by generating the hybrids through artificial cross-breeding experiments with the putative parents.

Only one specific name, normally the oldest published one, can legitimately be used for the hybrid population between two species, regardless of what varieties or subspecies may have been involved in the parentage. Thus the classification of the parents can affect the choice of names for their hybrid offspring. For example, if *P. perfoliatus* ssp. *richardsonii* is considered as a species (*P. richardsonii* (Bennett) Rydberg) distinct from *P. perfoliatus* L., the hybrids between it and *P. gramineus* L. are properly called *P.* x *hagstroemii* Bennett. If, however, *richardsonii* is combined with *P. perfoliatus* as ssp. *richardsonii* (Bennett) Hulten, then *P.* x *nitens* Weber, the oldest name given to a hybrid between any form of *P. perfoliatus* and *P. gramineus*, is applicable to all such hybrids, including those from ssp. *richardsonii*.

Many hybrids that have been identified in Europe (Dandy and Taylor 1938–1942) among species of circumboreal distributions are of potential interest here. The following hybrid combinations between parent species occurring in British Columbia have been recorded. Where specific names have been applied to the hybrids, they are included (Stace 1997). Asterisks indicate those that have been found to date in this province.

Potamogeton alpinus x *crispus* = *P.* x *olivaceus* Baagoe *ex* G. Fischer
P. alpinus x *gramineus* = *P.* x *nericius* Hagstroem
P. alpinus x *nodosus* = *P.* x *subobtusus* Hagstroem
P. alpinus x *perfoliatus* = *P.* x *prussicus* Hagstroem
P. alpinus x *praelongus* = *P.* x *griffithii* Bennett
* *P. amplifolius* x *illinoensis* = *P.* x *scoliophyllus* Hagstroem
* *P. amplifolius* x *natans*
* *P. amplifolius* x *nodosus*
P. amplifolius x *perfoliatus* ssp. *richardsonii*
* *P. amplifolius* x *praelongus*
P. crispus x *friesii* = *P.* x *lintonii* Fryer
P. crispus x *praelongus* = *P.* x *undulatus* Wolfgang
P. epihydrus x *nodosus* = *P.* x *subsessilis* Hagstroem
P. filiformis x *pectinatus* = *P.* x *suecicus* Richter
* *P. filiformis* x *vaginatus* = *P.* x *fennicus* Hagstroem

* *Potamogeton friesii* X *zosteriformis*
P. gramineus X *illinoensis* = *P.* X *spathulaeformis* (Robbins) Morong
P. gramineus X *natans* = *P.* X *sparganifolius* Hagstroem
P. gramineus X *nodosus* = *P.* X *argutulus* Hagstroem
* *P. gramineus* X *perfoliatus* = *P.* X *nitens* Weber
* *P. gramineus* X *perfoliatus* ssp. *richardsonii* (as *P. richardsonii*)
= *P.* X *hagstroemii* Bennett
P. illinoensis X *nodosus*
P. illinoensis X *perfoliatus* ssp. *perfoliatus*
P. illinoensis X *perfoliatus* ssp. *richardsonii*
P. natans X *nodosus* = *P.* X *perplexus* Bennett
P. nodosus X *perfoliatus* ssp. *richardsonii* = *P.* X *rectifolius* Bennett
* *P. pectinatus* X *vaginatus* = *P. bottnicus* Hagstroem
* *P. perfoliatus* X *praelongus* = *P.* X *cognatus* Ascherson and Graebner

Key to Species of *Potamogeton*

This key is modified from that by Hitchcock et al. (1969), using externally visible structural characters. For extensively eroded or otherwise mutilated material, it may be necessary to use less obviously visible characters, such as those of the internal stem anatomy, as seen by microscopic examination of cross-sections of the stems. A good key using these characters is that contributed by A. Ceska in Douglas et al. 1994.

1a. Sheaths adnate to leaf bases, at least of submerged leaves. Submerged leaves linear. 2
1b. Sheaths free of leaf bases. 6
2a. Leaves all alike, submerged, finely serrulate-margined, crowded in two ranks, stiff, over 3 mm wide. Flower spikes in terminal groups. ....*P. robbinsii*
2b. Leaves entire, 1 – 4 mm wide, not crowded. 3
3a. Adnate part of sheath of submerged leaf 5 mm long or less, shorter than free part. Elliptic floating leaves often present.......*P. diversifolius*
3b. Adnate part of sheath longer, at least as long as free part. Leaves filiform to linear. Floating leaves absent. 4
4a. Sheaths on main stem dilated, firm, at least some subtending two, three or more branches; the margins separate to base; free tips (ligules) relatively short. Leaf tip blunt or rounded. Achene 3 – 4 mm long; the stigma sessile. Flower whorls evenly spaced along spike. ....*P. vaginatus*
4b. Sheaths slender, subtending one or sometimes two branches on main stem. Ligules relatively longer. Leaf tip acute to obtuse. Flower whorls spaced more widely below than above....................5

5a. Sheath margins connate below when young, forming a tubular sheath often later ruptured. Leaf tip rounded to obtuse, seldom acute. Achene 2 – 3 mm long; the stigma sessile. ....*P. filiformis*
5b. Sheath margins free to base from the first, though often overlapping. Leaf tip acute. Achene 3 – 4.5 mm long, beaked. ....*P. pectinatus*
6a. Leaves all submerged and alike in form: oblong, 3 – 10 mm wide, with margins notably undulate and dentate. Stem flat. Achene with beak 2 – 3 mm long. ....*P. crispus*
6b. Leaves various, but not both undulate and dentate. Achene short-beaked or beakless. 7
7a. Stem flat, nearly as wide as leaves. Leaves all submerged, linear, with many parallel longitudinal fibres in addition to veins. Carpel one per flower. Achene keeled and beaked. ....*P. zosteriformis*
7b. Stem terete or moderately flat. Leaves various, but without fibres in addition to veins. Carpels four per flower. 8

8a. Leaves all submerged, cordate and usually clasping the stem, usually over 10 mm wide. 9
8b. Leaves not cordate or clasping at base. 10
9a. Stem unbranched or sparingly branched. Leaves usually less than 10 cm long, often recurving, acute or rounded at tips. Sheaths usually breaking down into fibres. Peduncle 2 – 10 cm (rarely more) long. Achene 3 – 3.5 mm long........ *P. perfoliatus*
9b. Stem profusely branched above, often zigzag. Leaves of various sizes, bundles of small leaves in axils of leaves 10 – 20 cm long, rounded and hooded at tips. Sheaths persisting whole. Peduncle 10 – 20 cm long. Achenes 4 – 8 mm long........ *P. praelongus*
10a. Stem with paired translucent glands at nodes. All leaves submerged and linear. 11
10b. Stem without glands. Leaves various. 14
11a. Sheaths white, fibrous, breaking down into persistent fibres. Stem flat........ 12
11b. Sheaths membranous, colourless, translucent, non-fibrous or nearly so, often caducous. 13

12a. Leaves thin, relatively flaccid, translucent, with five to seven longitudinal veins, and blunt or rounded tips........ *P. friesii*
12b. Leaves stiff, opaque, usually three-veined, with acute tips. Margins often becoming rolled under. .... *P. strictifolius*
13a. Leaves 2 – 4 mm wide, rounded at tips. Achenes 3.5 mm or more long........ *P. obtusifolius*
13b. Leaves 1 – 1.5 mm wide, acute at tips. Achenes 2 – 3 mm long. .... *P. pusillus*
14a. Leaves all alike, submerged, small, linear, less than 2 mm wide or 7 cm long. 15
14b. Leaves often diverse, and either longer or wider, or both. Floating leaves usually present. 16
15a. Sheath margins connate when young. Achene prominently keeled and beaked. Plant rhizomatous........ *P. foliosus*
15b. Sheath margins usually free from the first. Achenes keel-less, short-beaked. Plant not rhizomatous........ *P. pusillus*
16a. Submerged leaves narrowly linear, up to 2 mm wide........ 17
16b. Submerged leaves tapering to ends, 3 mm or more wide. 18
17a. Stem 1 – 2 mm thick, unbranched or sparingly branched. Floating leaf 4 – 12 cm long by 2 – 6 cm wide, commonly cordate at base and rounded at tip, on a petiole 1 – 2 mm thick. Submerged leaf stiff, 10 – 20 cm long........ *P. natans*
17b. Stem 0.5 – 1 mm thick, branched. Floating leaf 2 – 5 cm long by 0.5 – 2 cm wide, acute at base and tip, on a petiole 0.5 – 1 mm thick. Submerged leaf flaccid, 2 – 13 cm long........ *P. oakesianus*

18a. Submerged leaves sessile..19
18b. Submerged leaves on petioles 0.5 – 4 cm long...................................21
19a. Floating leaves subsessile or very short-petioled. Submerged leaves seven- to nine-veined, tapering to base and apex. Plant commonly reddish-tinged..*P. alpinus*
19b. Floating leaves with petioles usually at least 2 cm long....................20
20a. Submerged leaves lanceolate, usually less than 10 cm long, often crowded on densely branched lower stems, with midvein less than 1 mm wide..*P. gramineus*
20b. Submerged leaves linear, 10 cm or more long, on an unbranched or sparingly branched stem. Midvein conspicuous, 2 mm or more wide, including the adjacent bands of air channels (lacunae)......*P. epihydrus*

21a. Submerged leaves broadly elliptic, up to 7.5 cm wide, but often folded lengthwise and falcate-recurved, entire-margined. ....*P. amplifolius*
21b. Submerged leaves not folded or falcate-recurved, with minute marginal teeth (difficult to see). 22
22a. Submerged leaves mucronate at tips, 2 – 5 cm wide, on petioles 0.5 – 2 cm long. Floating leaves, if present, on petioles usually shorter than blades. Transitional leaves often present. ......*P. illinoensis*
22b. Submerged leaves acuminate, 0.5 – 2 cm wide, on petioles usually over 2 cm long. Floating leaves on petioles at least as long as blades. ....*P. nodosus*

Potamogeton alpinus Balbis — Alpine Pondweed

Generally reddish-tinged rhizomatous perennial with unbranched or sparingly branched stems up to a metre or more long, but usually shorter.

Submerged leaves lanceolate to narrowly elliptic, tapering to a sessile base and acute to rounded apex, 7 – 20 cm long by about a tenth to a fifth as wide, usually seven-veined, thin and translucent; with broad, thin, free-margined ligular sheaths quite free of the leaf bases. Floating leaves (often absent) obovate, broader than the submerged ones, relatively thick and firm, and with more veins; the blades up to 6 cm long by 2.5 cm wide, tapering to short petioles. Transitional leaves often present, even when floating leaves absent.

Peduncle red, about as thick as the stem or slightly thicker. Spike with five to nine whorls of flowers, up to 3.5 cm long at flowering time, elongating somewhat in fruit. Achene 3 – 4 mm long including the curved terminal stylar beak, with a short dorsal keel, orange to reddish when ripe. 2n = 52. Figure 23.

Key to Varieties

1a. Submerged leaves lanceolate, eight to ten times as long as their width, with obtuse to acute apices. ....var. *tenuifolius*

1b. Submerged leaves more elliptic, less than eight times as long as their width, the apices rounded, sometimes slightly hooded. ....var. *subellipticus*

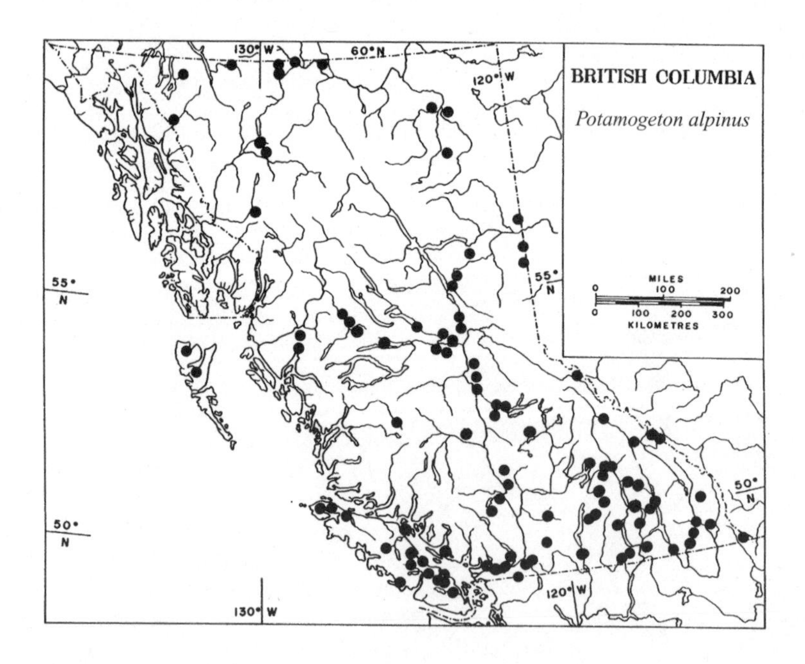

Potamogeton alpinus is circumboreal in distribution. Variety *alpinus*, the typical form of the species, which is distinguished from our varieties by its achene with a straighter beak and more convex ventral margin, occurs from Europe to central Asia. Variety *tenuifolius* (Rafinesque) Ogden occurs from eastern Siberia across North America to Greenland, and var. *subellipticus* (Fernald) Ogden across North America. Generally northern and alpine or subalpine in range, *Potamogeton alpinus* is found as far south as Colorado and California at high altitudes. It grows in shallow, cold ponds and lakes, 0.5 – 1.5 (– 3.5) metres deep, commonly where the bottom is non-calcareous or has deposits high in organic matter.

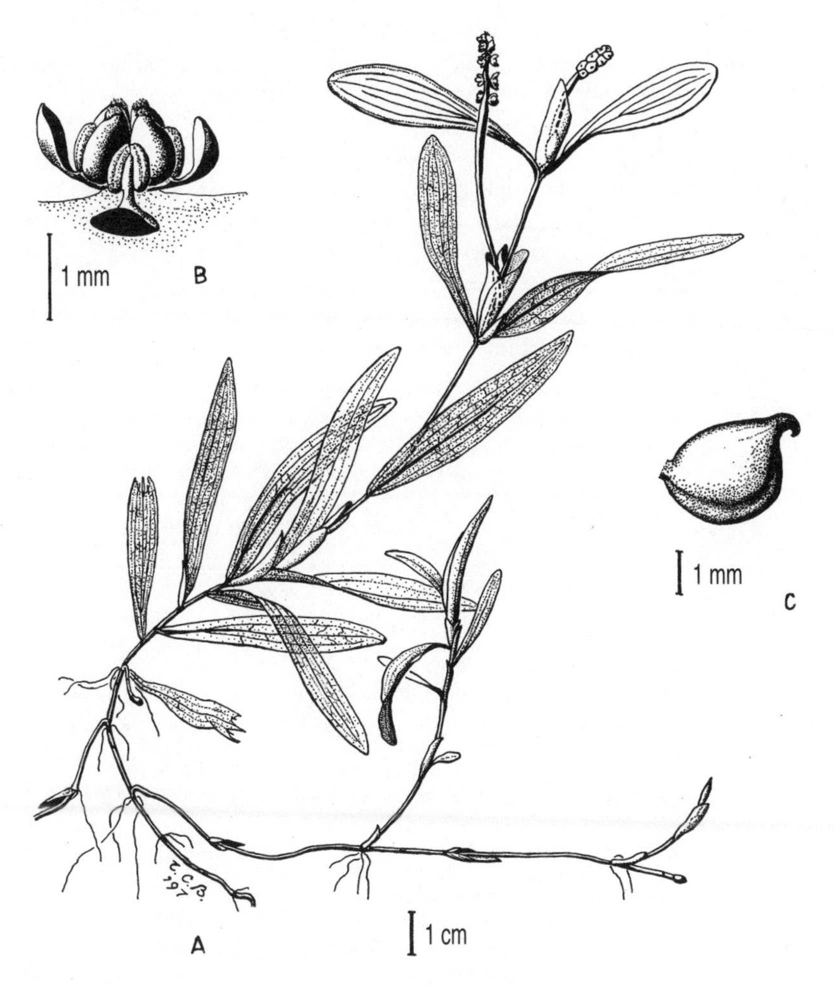

Figure 23. *Potamogeton alpinus* var. *subellipticus*: A, shallow-water plant; B, flower; C, fruit (achene).

Potamogeton amplifolius Tuckerman **Large-leafed Pondweed**

Coarse perennial, 1 – 5 metres or more long, with stems up to 3 mm thick, sparingly branched, from a long coarse rhizome.

Leaves of two kinds. The submerged leaves on short petioles usually 1 – 2 cm long, the blades up to 20 cm long by up to 7.5 cm wide, typically folded along the midvein and falcate-recurved as seen from the side, and loosely undulate, as the entire margin elongates more than the midvein; they have 19 – 45 longitudinal veins, are olive green and translucent, and the deeper leaves narrower, fewer-veined and straighter. The floating leaves on petioles longer than the elliptic blades, which are relatively thick and opaque, green and flat, 5 – 10 cm long by 2 – 4 cm wide; they have 25 – 45 longitudinal veins. Sheaths of all leaves are up to 10 cm long, free from the petioles, translucent, whitish and acute, and they have free margins.

Peduncles enlarged upward. Spikes densely flowered with up to 16 whorls of flowers, and 1 cm or more thick in fruit. Achene obovoid, 4 – 5 mm long, with a stout beak 0.5 – 1 mm long, tipped by a recurved stigma. Ripe achenes are orange to pinkish in colour. Dried achenes show prominent dorsal and lateral keels, which are not seen on fresh achenes. 2n = 52. Figure 24.

This species ranges discontinuously across North America, and from southern British Columbia southward to California in the west. It grows in deep clear water to 4 metres deep, or rarely to 6 metres in very clear water. It may show a high rate of spring growth. Plants over 3 metres long have been

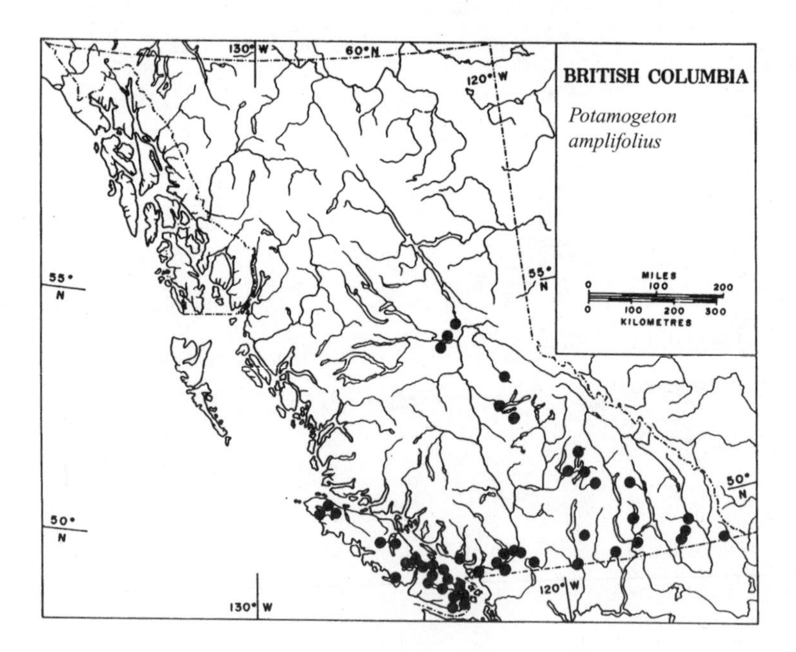

observed at the beginning of May in the Victoria area, leading to a suspicion that growth may start from the bottom of the lake before winter sets in.

Some plants in Thetis Lake, near Victoria, combine the coarse stems and green floating leaves of *P. amplifolius* with the sharply two-keeled sheaths and lowest leaves reduced to bladeless petioles of *P. natans*. The upper submerged leaves have long (up to 12 cm) petioles, and long, narrow (to 12 by 1.5 cm) in-rolled blades. The pollen grains are nearly all empty, and no mature fruit has been found. This plant is believed to be the previously unrecorded putative hybrid *P. amplifolius* X *natans*. Both the parent species occur nearby.

Figure 24. *Potamogeton amplifolius*: A, upper part of plant with flowering shoot and submerged sterile shoot: B, flower; C, achenes.

Potamogeton crispus L. **Crisped Pondweed**

Extensively growing, freely branching, leafy-stemmed plant with flat, wing-edged stems. Rhizomes are produced that may overwinter in mild conditions, but the usual method of overwintering vegetative propagation is with winter buds. Each such bud germinates in spring to produce several shoots and rhizomes from small buds in the axils of its reduced leaves.

Leaves all immersed, sessile, oblong, up to 10 cm long by 1 cm wide, stiff and rather cartilaginous in texture, translucent, with three main veins, and notably dentate and undulate margins. Winter grown leaves may not have undulate margins. The first node on a branch produces a sheath without an associated leaf blade. Sheaths delicately membranous, free of the leaf bases, with overlapping margins, closely surrounding the stem, soon disintegrating.

Peduncles curved, somewhat dilated above, about twice as long as the 1 – 3 cm long spikes. Flowers with long-clawed tepals, and carpels tapering into prominent stylar beaks. Achenes 4 – 6 mm long, including a beak 2 – 3 mm long, and with a strongly toothed dorsal keel. 2n = 52. Figure 25.

Potamogeton crispus is a species of European origin, commonly grown in aquaria, and discarded into ditches, etc. It has become locally an aggressive weed in lakes and in deeper ditches and canals.

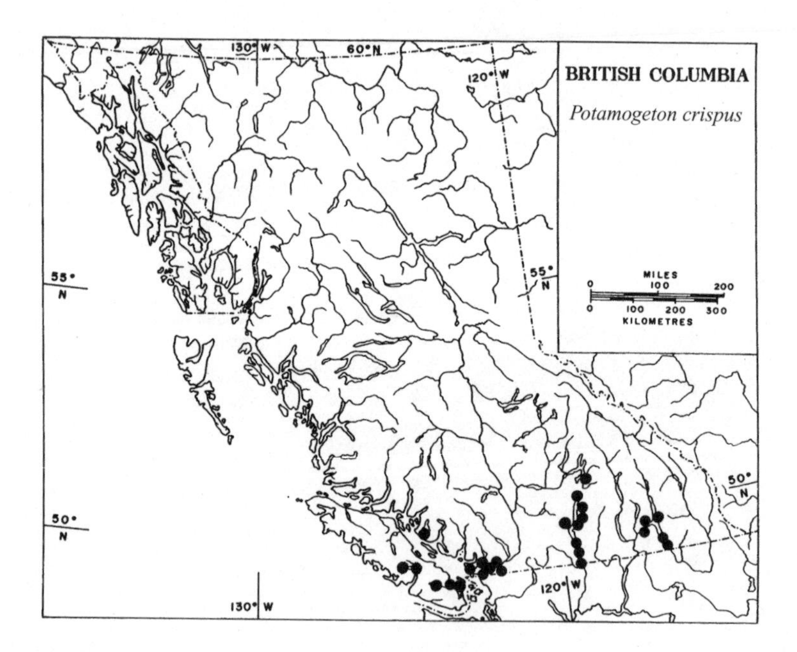

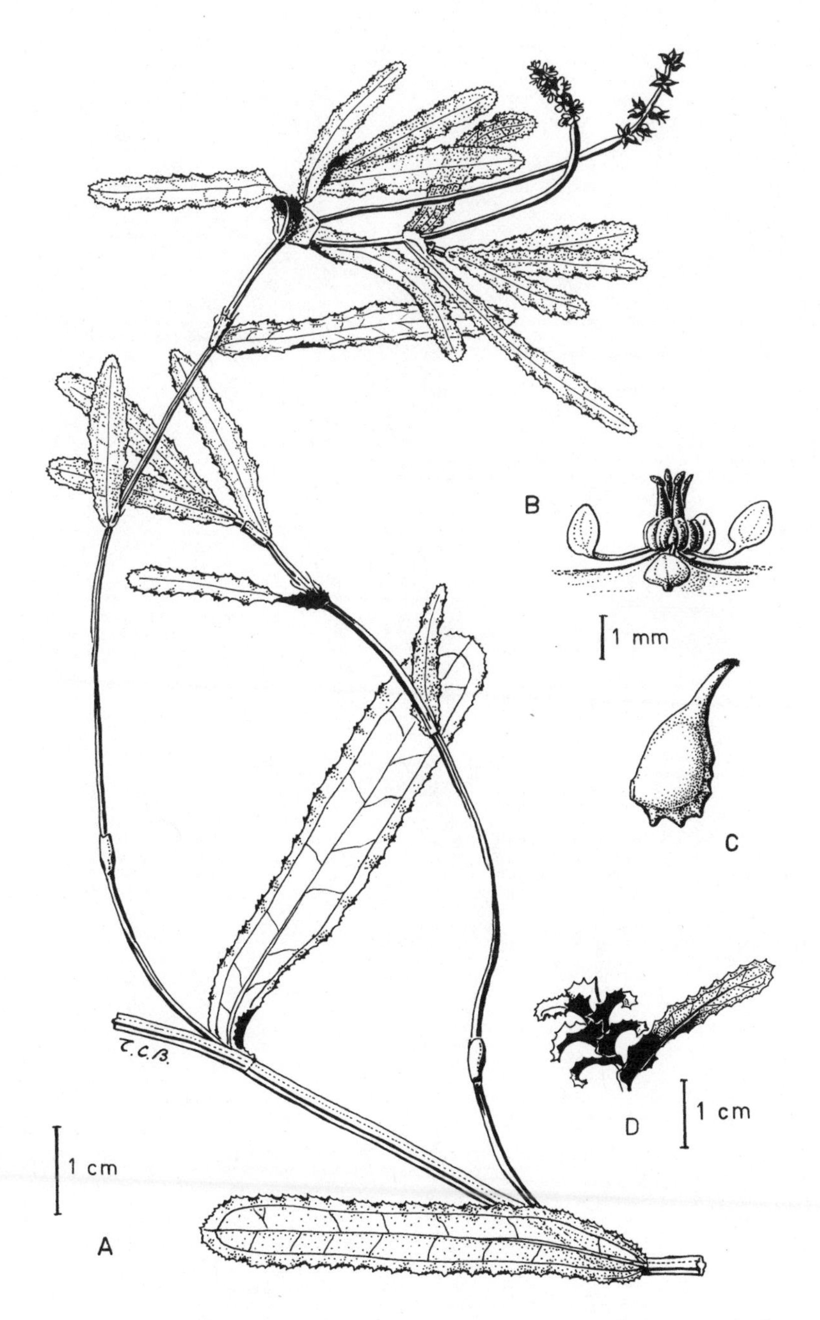

Figure 25. *Potamogeton crispus*: A, plant; B, flower; C, achene; D, winter bud.

Figure 26. *Potamogeton diversifolius*: A, plant; B, achene.

Potamogeton diversifolius Rafinesque — Diverse-leaved Pondweed

Freely branched plant up to 1.5 metres long, with a slender terete stem; propagating vegetatively by winter buds produced from upper leaf axils in autumn.

Leaves of two kinds. Submerged leaves sessile, linear, up to 4 cm long by 1.5 mm wide, one- to three-veined, acute to obtuse at tips, adnate at base to the lower third of the sheath. Floating leaves with elliptic blades up to 3 cm long, rounded at apex, on petioles about half as long which are free from or barely adnate to the sheaths. Sheaths up to 1 cm long by submerged leaves, up to 3 cm long by floating leaves, open to the base on the side opposite the leaf base.

Peduncles and spikes very short, appearing beneath the water as well as above; the lower spikes with one to three whorls of flowers, the upper, emergent spikes with up to six whorls. Achene compressed, 1 – 1.8 mm long, distinctly keeled on the dorsal margin, with a very short straight beak. Embryo coiled into more than a complete turn, discernible externally. Figure 26.

Potamogeton diversifolius is a species of southern distribution, ranging across the United States from Idaho eastward, northward into Alberta, and southward into northern Mexico. This species was reported as occurring in British Columbia by Hitchcock (1969). But the only material from this province that I know of (reported in the first edition of this book as from Steelhead near Mission) has since then been re-identified as *P. oakesianus*. At present, I know of no certain record of *P. diversifolius* from British Columbia. The description is retained here, since it may yet be found in this province, possibly in the southeastern interior. The illustration in figure 26 is based on material from eastern Canada.

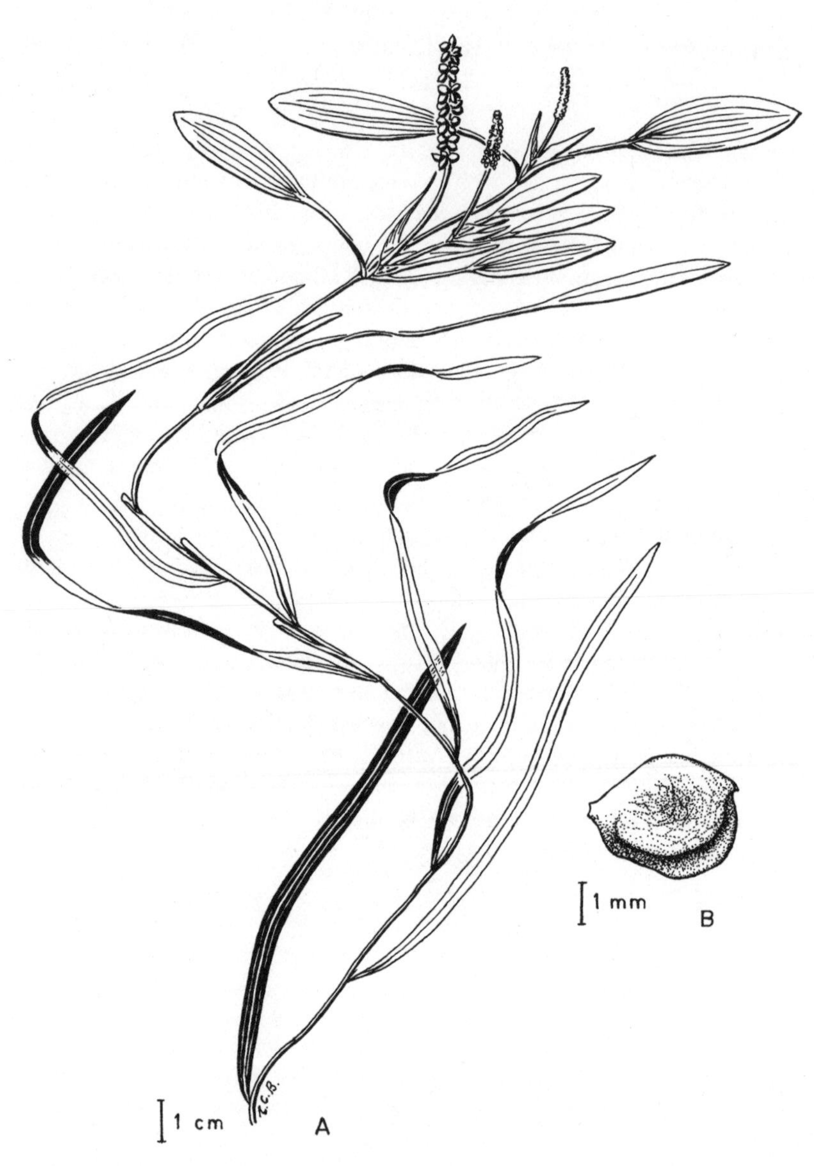

Figure 27. *Potomageton epihydrus* var. *ramosus*: A plant; B, achene.

Potamogeton epihydrus Rafinesque **Ribbon-leafed Pondweed**

***P. epihydrus* var. *nuttallii* (Chamisso and Schlechtendal) Fernald = var. *ramosus* (Peck) House**

Rhizomatous perennial, the stem moderately compressed and sparingly branched or unbranched. Sometimes reproducing by large winter buds.

Leaves of two kinds. Submerged leaves linear and ribbonlike, 4 – 25 cm long by 2 – 10 mm wide, shortly tapering to cuneate sessile bases and acute tips, flaccid, with 3 – 13 longitudinal veins, the midvein bordered by conspicuous bands of loosely cellular lacunate tissue. Floating leaves with narrowly obovate to elliptic, leathery blades, 3 – 8 cm long by 1.5 – 3.5 cm wide, with up to 40 veins, rounded to obtuse at tips and tapering to cuneate bases, on petioles from shorter than to rather longer than the blades. Transitional leaves also occur. Submerged leaves and their sheaths on fertile stems often decadent by flowering time. Sheaths delicately membranous, free from leaf bases, translucent, up to 4 cm long, obtuse on nodes bearing submerged leaves, somewhat longer and attenuate on nodes bearing floating leaves.

Peduncles up to 16 cm long, as thick as stem or somewhat thicker, not tapering. Spike 2 – 4 cm long, with 7 – 12 whorls or pairs of flowers. Spikes may arise at nodes bearing submerged leaves, especially in deep clear water. Achene nearly circular and flat, with a pronounced dorsal keel, in dried material at least, and the beak, if any, inconspicuous. 2n = 26. Figure 27.

Key to Varieties

1a. Submerged leaves 5 – 10 mm wide with 7 – 13 veins. Floating leaves broadly elliptic, up to 35 mm wide with 20 – 40 veins. .... var. *epihydrus*

1b. Submerged leaves 2 – 8 mm wide with 3 – 7 veins. Floating leaves narrowly obovate or oblanceolate, 5 – 25 mm wide, with 7 – 33 veins.. var. *ramosus*

These varieties occupy much the same range generally, and intergrade completely. Most British Columbian material appears to be var. *ramosus* (Peck) House (var. *nuttallii* (Chamisso and Schlechtendal) Fernald).

This transcontinental species ranges from Alaska to Labrador, and southward over all but the southernmost United States, with an outlying range in Scotland. It is found in clear water 0.5 – 3 metres deep, often with a silty or sandy bottom.

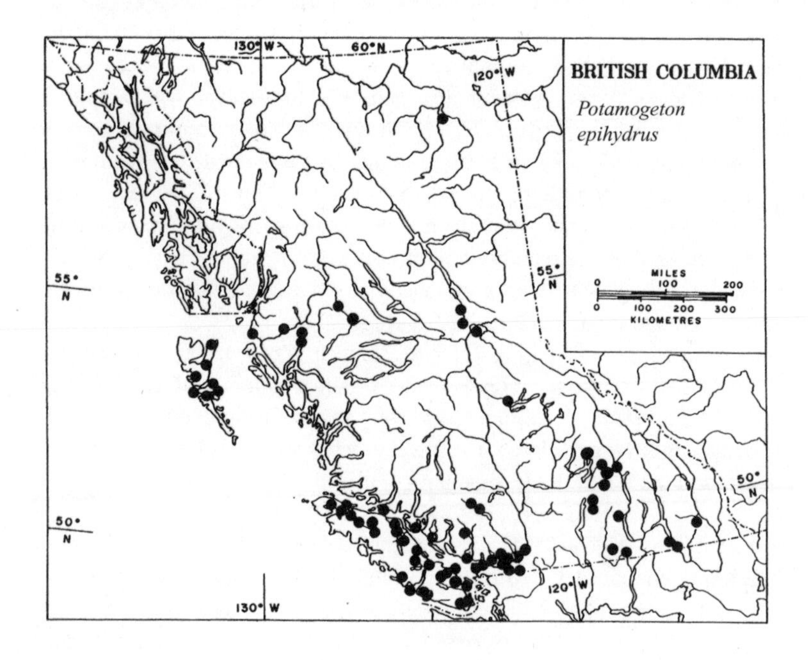

Potamogeton filiformis Persoon — Thread-leafed Pondweed

P. marinus **L. var.** ***macounii*** **Morong in Macoun =** ***P. filiformis*** **var.** ***macounii*** **(Morong in Macoun) Morong =** ***P. filiformis*** **var.** ***occidentalis*** **Robbins in Watson =** ***P.*** **x** ***fennicus*** **Hagstroem (*****P. filiformis*** **x** ***vaginatus*****)**
P. borealis **Rafinesque =** ***P. filiformis*** **var.** ***borealis*** **(Rafinesque) St John =** ***P. filiformis*** **var.** ***alpinus*** **Blytt**
P. interior **Rydberg =** ***P. filiformis*** **var.** ***occidentalis*** **=** ***P.*** **x** ***fennicus***
Stuckenia filiformis **(Persoon) Borner**
Coleogeton filiformis **(Persoon) Les and Haynes**
C. filiformis **ssp.** ***alpinus*** **(Blytt) Les and Haynes =** ***P. filiformis*** **var.** ***alpinus***
C. filiformis **ssp.** ***occidentalis*** **(Robbins) Les and Haynes =** ***P. filiformis*** **var.** ***occidentalis*** **=** ***P.*** **x** ***fennicus*** **Hagstroem**

Entirely submerged perennial, overwintering mainly as small tubers terminating the slender buried rhizome. Stem filiform, terete, freely branched, sometimes with more than one branch at a node, the plant often having the aspect of a small dense brush.

Leaves filiform, up to 10 cm long by 0.5 – 2 mm wide, tapering to an obtuse or rounded tip, one- to three-veined, often brownish in colour. Sheath slender, membranous, adnate to the leaf base for more than half of its length, tubular below with connate margins, at least when young, though commonly ruptured and split to base later by growth of an axillary branch; the ligule from a sixth to nearly half the length of the whole sheath.

Peduncle filiform, up to 15 cm long, bearing an open spike of two to eight whorls of flowers spaced at varying distances apart, the lowest whorl often remote from the others. Achene obovoid, 2 – 3 mm long, beakless, with a sessile, apical or subapical stigma, light brown when ripe. Fresh achenes show no distinct keel, but dried material may display keel-like wrinkles. 2n = 78. Figure 28.

As the above list of synonyms attests, this is a confusingly diverse species – several varieties or subspecies, some with overlapping character ranges, have been described. Those that have been reported for British Columbia, or may be expected here, are distinguished by the key on page 111.

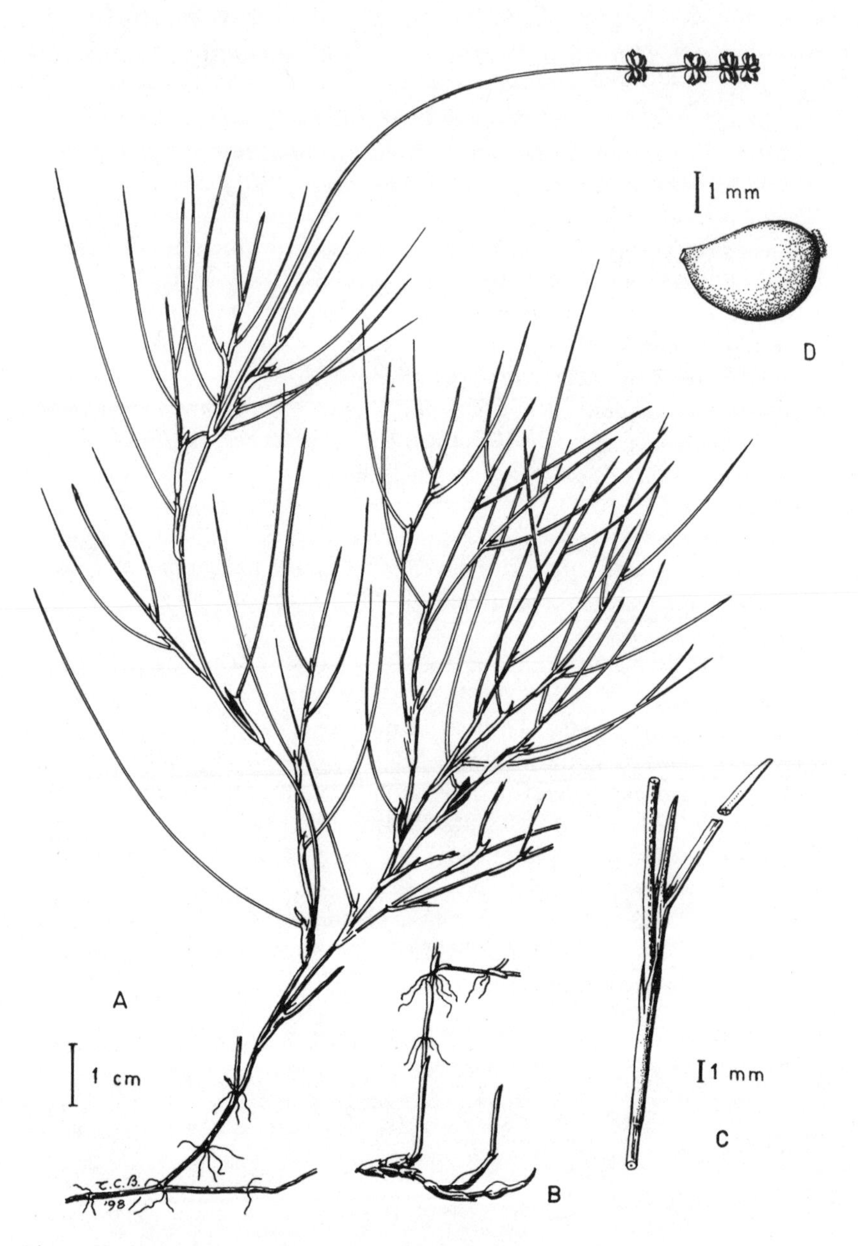

Figure 28. *Potamogeton filiformis*: A, fruiting plant; B, rhizome with tubers; C, sheath with ligule, leaf base and apex; D, achene.

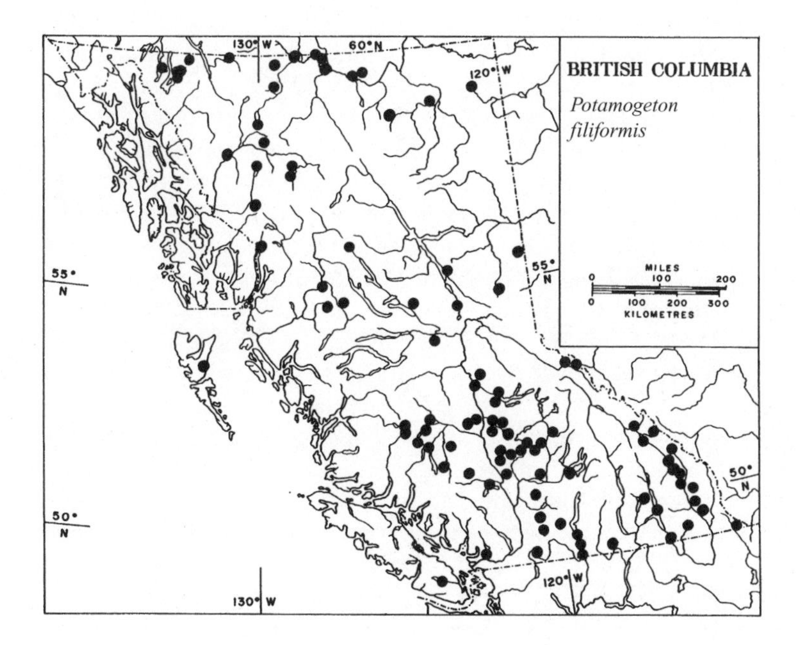

Key to Varieties

1a. Mature spike elongate, the whorls, especially the lowest, separated by 1 – 3 cm, the upper whorls closer. Leaves 0.3 – 1.2 mm wide. ....var. *filiformis*

1b. Mature spike shorter, with closely spaced whorls, the lower up to 0.7 – 1.0 cm apart. 2

2a. Leaves filiform, mostly less than 0.7 mm wide. All sheaths close-fitting, scarcely wider than stems. Plant commonly of short stature: 5 – 20 cm tall...var. *alpinus*

2b. Leaves 0.5 – 2 mm wide. Lower sheaths somewhat dilated. Taller plant, 20 – 60 cm tall. ....var. *occidentalis* (= *P.* x *fennicus*)

Potamogeton filiformis ranges over most of the northern hemisphere, extending southward to Africa and Australia. It has been found over most of British Columbia, including recently, the Queen Charlotte Islands (Lomer and Douglas 1999). It is found in still or slow-moving, often calcareous water, commonly in shallower water (15 – 100 cm) than that occupied by the related *P. pectinatus*. The stature of the plant is strongly influenced by the depth of the water in which it grows. Variety *alpinus* in particular, may be found in a very dwarfed form in water 20 cm or less in depth.

At present the collection at the Royal British Columbia Museum contains no specimens from British Columbia of var. *filiformis*, which is primarily an Old World variety. But C.B. Hellquist has annotated a specimen from Slim's

River, Yukon (60° 54' N, 138° 37' W) as this variety. A specimen identified as ssp. *alpinus* from Lake Eddontenajon (57° 43' N, 129° 59' W) in northwestern British Columbia has an inflorescence with the lower whorls spaced 13 – 20 mm apart, and the leaves (0.3 –) 0.4 (– 0.6) mm wide; thus appearing intermediate between the above two varieties. Variety *filiformis* is included here in the expectation that it will in time be found in this province in its typical form.

Variety *occidentalis* (Robbins in Watson) Morong, in which Cronquist et al. (1977) include var. *macounii* (Morong in Macoun) Morong, has been recorded from widely scattered locations within the range of the species in British Columbia. Hellquist and Crow (1980) have drawn attention to a feature of this variety that is not mentioned by other authors; namely, that this variety is characterized by the presence of inflated sheaths on the lower stems of young plants, which thus can readily be confused with *P. vaginatus*. This feature appears to reflect a genetic origin by introgression of genes of *P. vaginatus* into *P. filiformis* through past incidents of hybridization. Thus, var. *occidentalis* can be looked on as a member of the hybrid population of *P.* x *fennicus* Hagstroem, which is known to exist between *P. filiformis* and *P. vaginatus*. British Columbian records of this entity are shown in the map below.

Potamogeton filiformis flowers are water-pollinated, either by floating pollen (Sculthorpe 1967) or by submerged pollen (Dandy 1980 c, p. 10). Flowering spikes are often seen beneath the water surface.

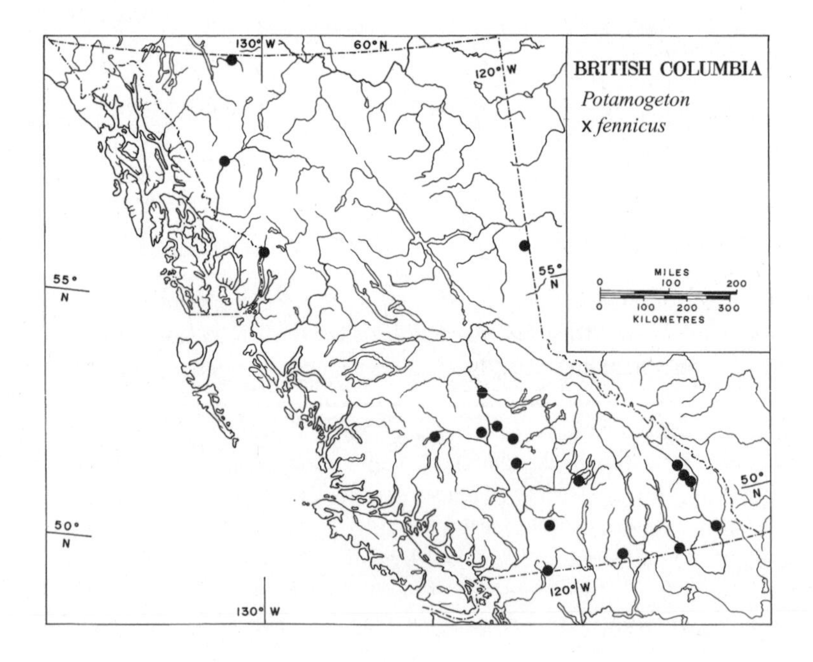

Potamogeton foliosus Rafinesque Close-leaved Pondweed

Submerged plant with filiform rhizomes. Stems up to 1 metre long, freely and divergently branching, slender, slightly flat, usually without paired glands at the nodes. Perennation can occur by winter buds produced on the ends of short lateral branches or in leaf axils.

Leaves all alike, sessile, narrowly linear, 2 – 10 cm long by 1 – 2.5 mm wide, acute-tipped, with 1 – 5 veins, the midvein usually bordered by air channels, at least near the base. Sheath free from the leaf base, with connate margins below when young, later ruptured by axillary branch growth, finally disintegrating.

Peduncles short, 1 – 3 cm long, the spikes 0.5 – 1 cm long, with two to four whorls of flowers initially crowded but later separated by elongation of the axis. Tepals 0.4 – 1 mm long. Achene about 2 mm long, with a conspicuous undulate dorsal keel and a beak up to 0.5 mm long. 2n = 28. Figure 29.

Key to Varieties

1a. Leaves 1.5 – 2.5 mm wide with three to five longitudinal veins. Stems sparingly branched. ....var. *foliosus*

1b. Leaves 1.5 mm or less wide with one to three veins. Stems freely branched. ....var. *macellus*

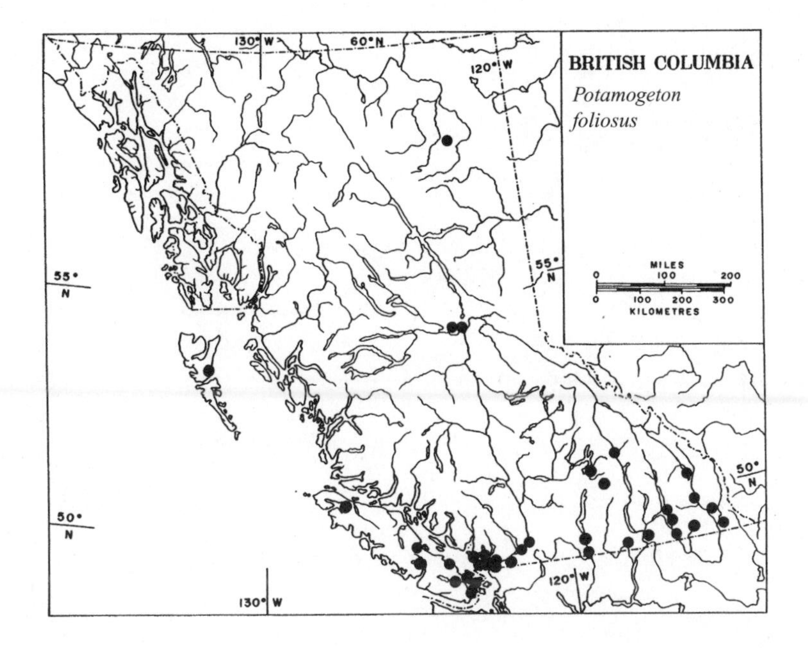

Potamogeton foliosus is widespread in North and Central America. Variety *foliosus* is generally more southern in range, its occurrence in British Columbia being uncertain. Variety *macellus* Fernald is the common variety in British Columbia. It has recently been found in the Queen Charlotte Islands (Lomer and Douglas 1999) and it probably occurs in northwestern British Columbia, since its northern range reaches the Yukon and Northwest Territories. It overlaps with var. *foliosus* in the United States.

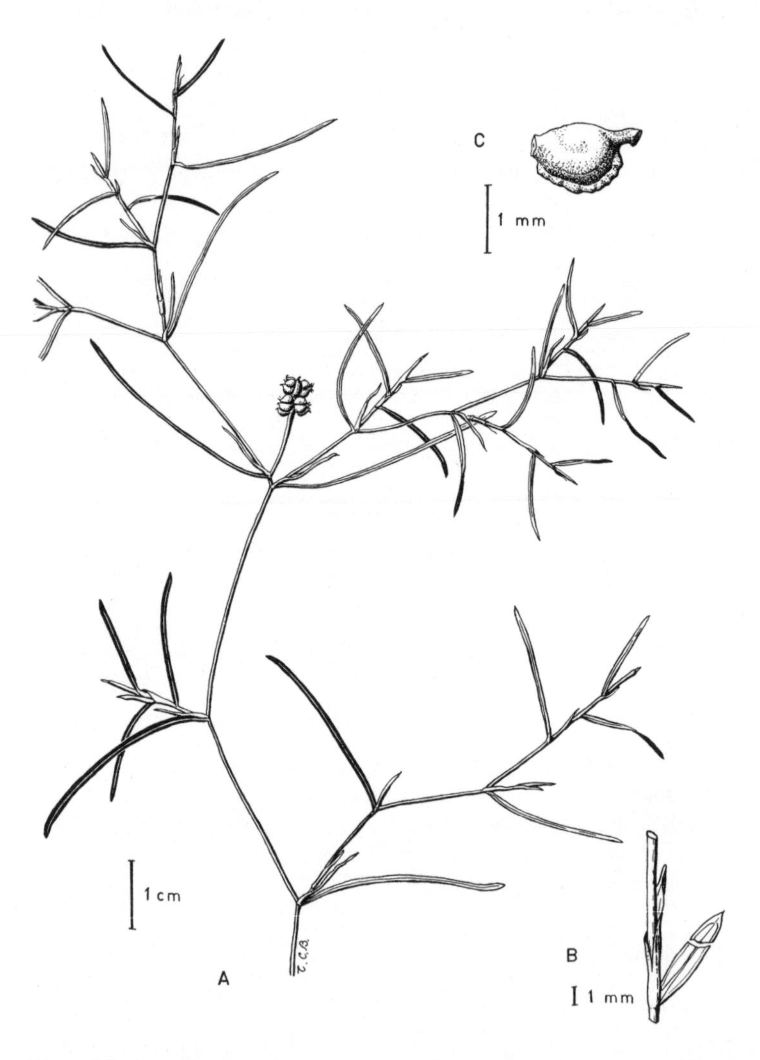

Figure 29. *Potamogeton foliosus* var. *macellus*: A, plant; B, sheath and ligule; C, achene.

Potamogeton friesii Ruprecht **Flat-stemmed Pondweed**

***P. compressus* L.?**
P. mucronatus* Schrader *nomen nudum

Submerged plants, sometimes with slender rhizomes, with freely branching, slender, flat stems up to 1 metre long, with paired yellowish translucent glands at the nodes. Perennating by winter buds of peculiar structure, which are borne on the ends of short axillary branchlets. These buds, which average about 2 – 2.5 cm long, bear two ranks of short stiff leaves, with their vascular strands stiffened with fibrous tissue at their bases. Their sheaths are divided fore and aft, and the resulting half-sheaths, white and fibrous, diverge in two ranks in a plane perpendicular to that containing the leaves.

Leaves all alike, sessile, linear, up to 8 cm long by 3 mm wide, rounded and mucronate at the tips, commonly with three to five longitudinal veins. Sheaths conspicuous, up to 18 mm long, white, fibrous, and opaque, with margins connate when young; finally shredding into the persisting fibres.

Peduncles flat, curved at base, up to 5 cm long, but commonly less. Spikes emergent at flowering time, 1 – 2 cm long, with two to four whorls of flowers. Achenes 2 – 2.5 mm long, with an obscure rounded dorsal keel, and a more or less curved terminal beak up to 0.6 mm long. 2n = 26. Figure 30.

A circumboreal species, *P. friesii* ranges from Alaska to Labrador, reaching the Arctic coast and Hudson Bay (Porsild and Cody 1980), and from there southward into the northern United States. It is widespread across British Columbia. It is found in fresh to brackish still water 0.5 – 1.5 (– 3.5) metres deep.

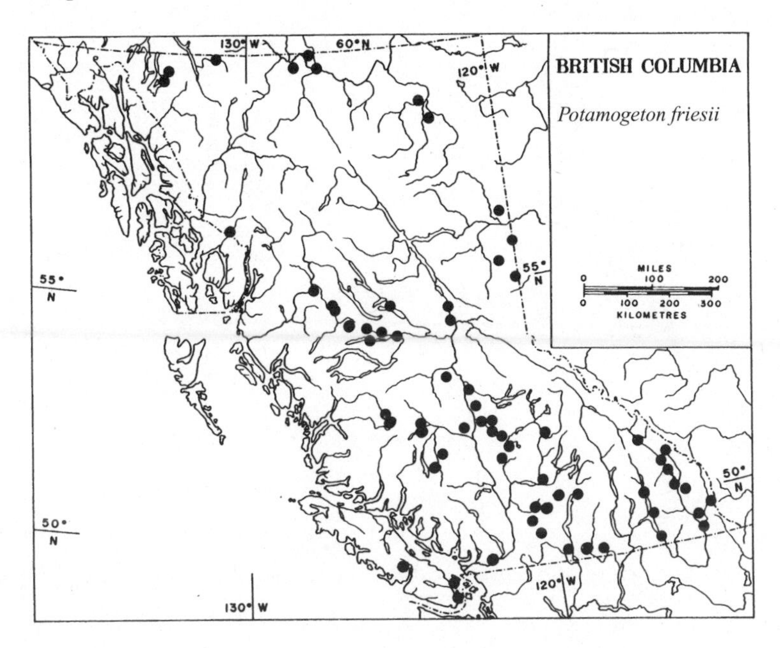

Specimens intermediate in character between this species and *P. zosteriformis* have been found at Elk Lake, near Victoria, and at Lac La Hache. They have the slender, flat but unwinged stem, not narrowed at the nodes, like *P. friesii*. The leaves are 6 – 12 cm long by 1.5 – 3 mm wide, have a few irregular interneural strands and acutish tips, and sheaths up to 3 cm long; flowers have one to three carpels, but no mature fruit. These plants are tentatively regarded as the hybrid *P. friesii* x *zosteriformis*.

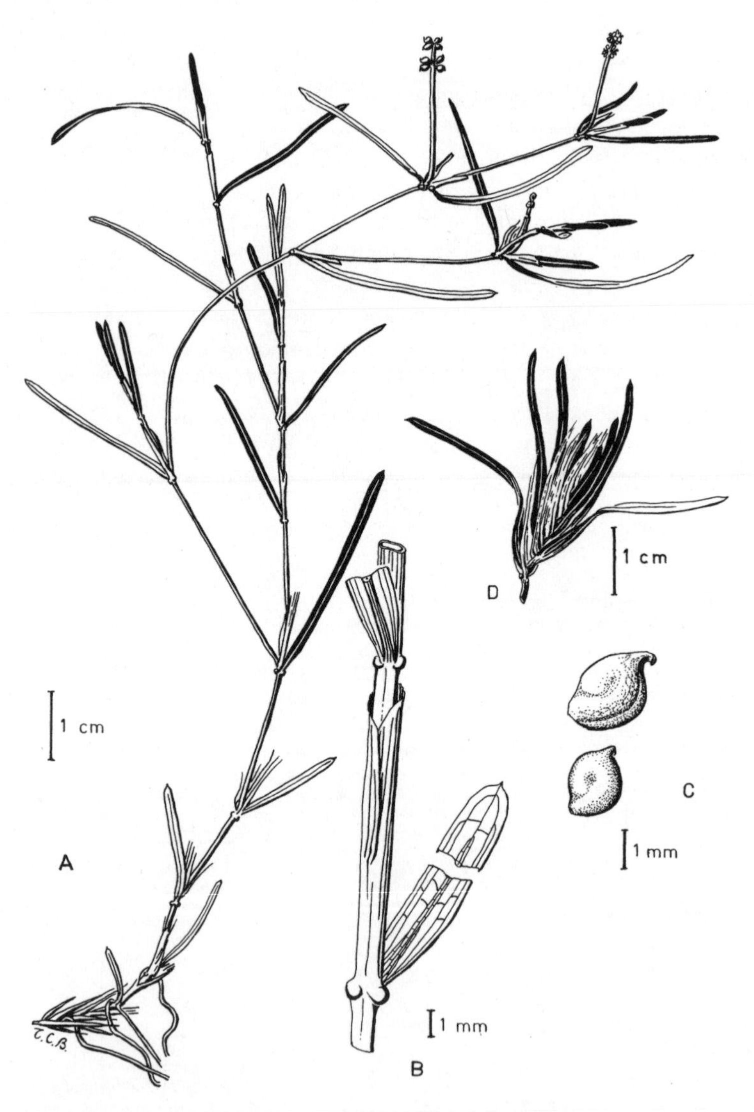

Figure 30. *Potamogeton friesii*: A, plant; B, detail of node, sheath and leaf; C, achenes; D, winter bud.

Potamogeton gramineus L. **Grassy Pondweed**
***P. heterophyllus* Schreber**

Our most variable species of this genus is *Potamogeton gramineus*, a perennial with slender rhizomes that produce elongate overwintering tubers in autumn. Stems varying from a few centimetres to over a metre long, according to situation, very slender, usually freely branching, occasionally with two branches in a leaf axil. The foliage is usually bright green, but varies to bronzy or occasionally reddish.

Leaves of two kinds. Submerged leaves sessile, lanceolate to linear, diverse in size on one plant, those on lateral branches smaller than the main stem leaves, which may be up to 13 cm long by a centimetre or more wide, tapering to bases and to acute or acuminate tips, 3 – 11 veined, with a narrow midvein, and with minute one-celled marginal teeth that are barely visible at 10x magnification. Floating leaves with elliptic blades up to 6 cm long by 1 – 3 cm wide, with up to 19 longitudinal veins, on petioles usually longer than the blades. Sheaths up to 3 cm long, open down one side.

Peduncles emergent, as thick as or thicker than the stems, and often thickened upward, of variable length up to 20 cm, according to situation. Spike dense, up to 3 cm long, with about 6 whorls or pairs of flowers. Achenes crowded, 2 – 3 mm long, rather flat, with a dorsal keel and a short straight beak. 2n = 52. Figure 31.

Key to Varieties

1a. Submerged leaves on main stem lanceolate to narrowly elliptic, up to 13 cm long by 3 – 15 mm wide, 5 – 10 times as long as their width, with five to eleven veins. 2

1b. Submerged leaves linear except for tapering ends, up to 6 cm long by 1 – 3 mm wide, 10 – 20 times as long as their width, with one to three veins. .... var. *myriophyllus*

2a. Submerged leaves on main stem 2 – 6.5 cm long by 3 – 8 mm wide, with five to seven veins. Stems normally profusely branched. .... var. *gramineus*

2b. Submerged leaves on main stem 3 – 13 cm long by 6 – 15 mm wide, with seven to eleven veins. Stem sparingly branched....... var. *maximus*

Widespread in temperate and subarctic regions of the northern hemisphere, this is a common species all over British Columbia. Variety *gramineus* is the common variety throughout this province. Variety *maximus* Morong *ex* Bennett occurs sporadically in the southern and central interior, usually in deeper water than the other varieties. Variety *myriophyllus* Robbins (= var. *graminifolius* Fries according to Fernald 1921) has so far been collected at only a few places on Vancouver Island.

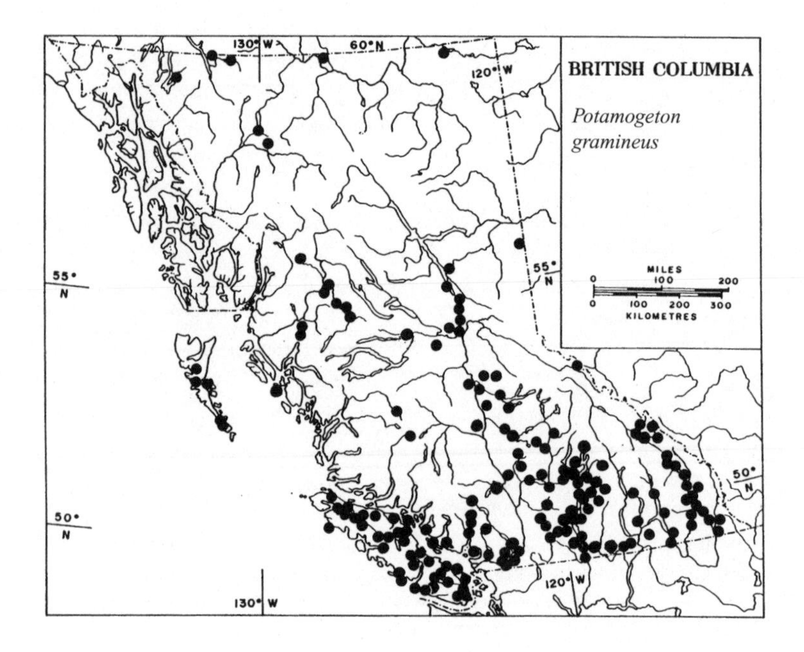

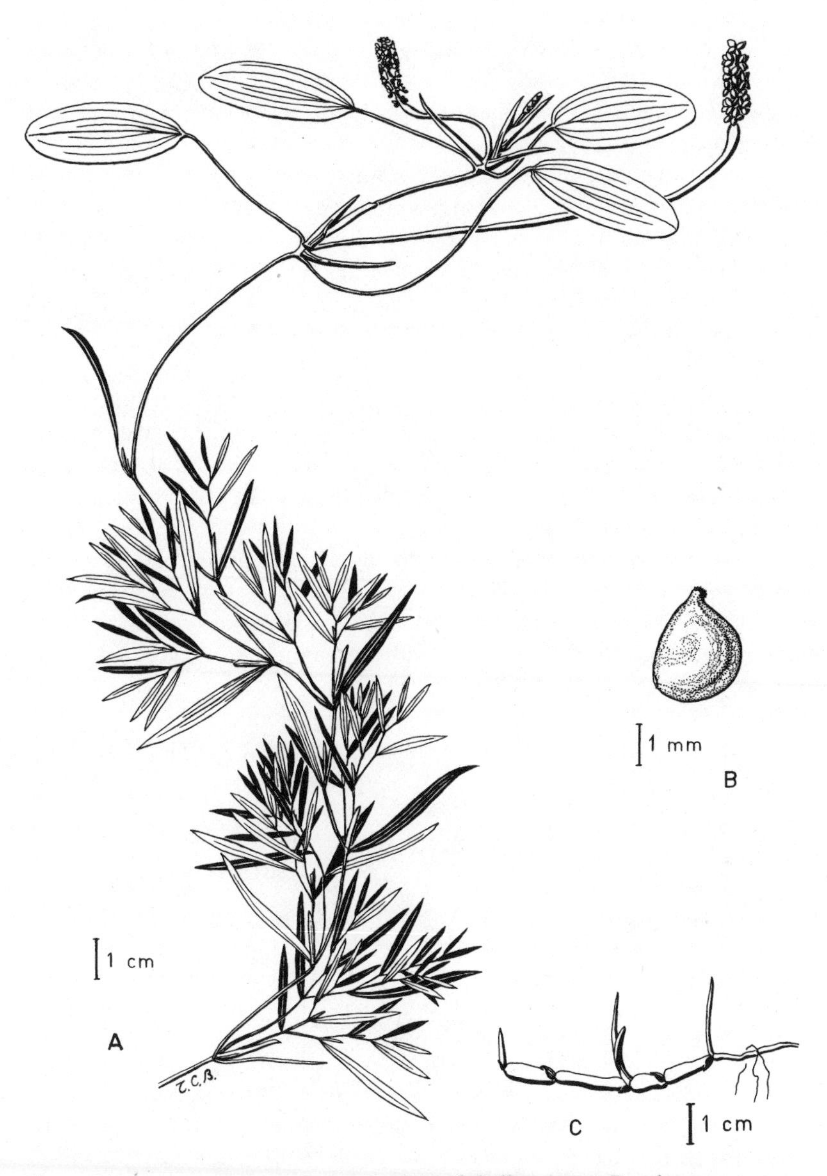

Figure 31. *Potamogeton gramineus*: A, plant of shallow water; B, achene; C, two tubers on rhizome.

Potamogeton illinoensis Morong **Illinois Pondweed**

Rhizomatous, long-stemmed plants up to over 3 metres long, with simple or branched stems.

Leaves of two kinds. Submerged leaves ample, rather shiny and translucent, sessile or on petioles up to 3 cm long, with blades up to over 20 cm long by 5 cm wide, lanceolate with cuneate bases and mucronate apices, with 9 – 19 longitudinal veins and conspicuous cross-veins, producing a netted appearance; the margins serrulate with minute, deciduous, one-celled teeth. Floating leaves, when present, on petioles shorter than the blades; the blades 6 – 12 cm long by 2 – 6 cm wide, thick and opaque, with 13 – 29 longitudinal veins. Transitional leaves commonly produced, similar in texture to typical submerged leaves, but relatively shorter and broader, with rather longer petioles and more broadly rounded (but still mucronate) apices. Sheaths prominent, up to 10 cm long, open to the base on the side opposite the leaf base, and commonly with two dorsal ridges.

Peduncle thicker than the stem, commonly extending well beyond the foliage, and emergent. Flower spike at first compact, 3 – 6 cm long, later opening out somewhat, with around 15 irregular whorls of flowers. Achene about 4 mm long, including the 0.5 mm beak, with a prominent dorsal keel and sometimes lateral keels. 2n = 104. Figure 32.

Potamogeton illinoensis is widespread in the United States (except the southeast) and southern Canada. In British Columbia, it is widespread in the

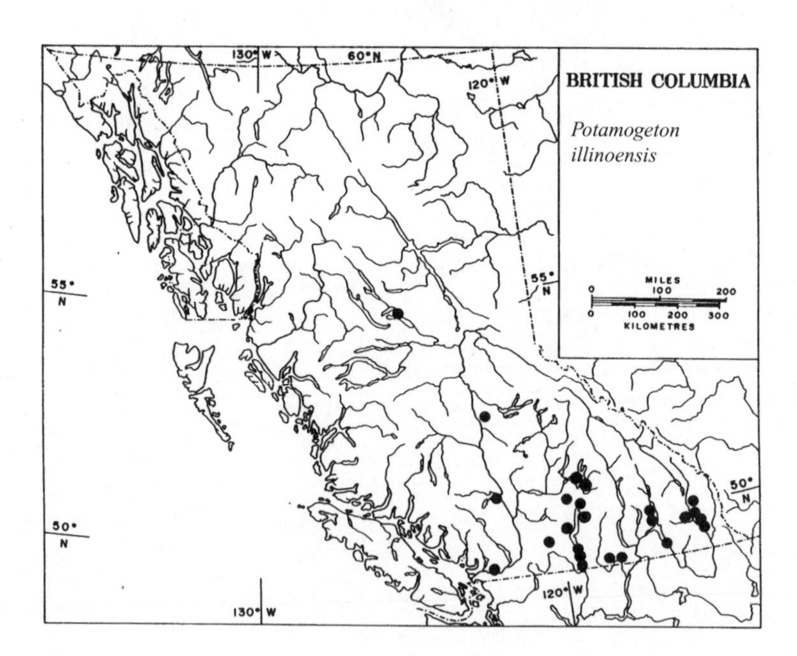

interior south of 55°N latitude, with a concentration of records toward the southeastern interior, but not recorded on the coast. It is found in deep still water of lakes and slow rivers, in water to 3 metres or more deep.

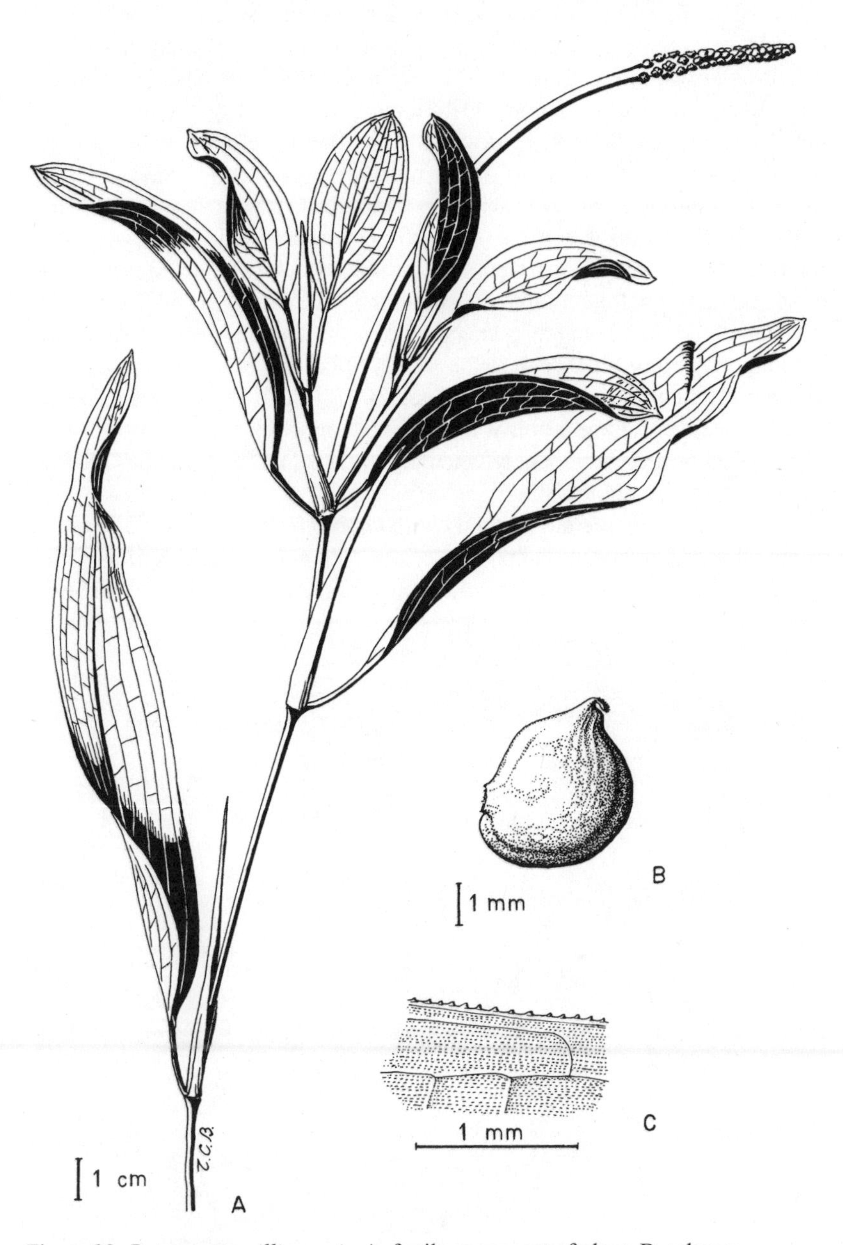

Figure 32. *Potamogeton illinoensis*: A, fertile upper part of plant; B, achene; C, detail of leaf margin.

Potamogeton natans L. **Broad-leafed Pondweed**

Rhizomatous perennial with strong, terete, green to brownish, usually unbranched stems up to 3 metres long and 2 mm thick, but usually shorter and thinner; and a rhizome that commonly produces overwintering tubers.

Leaves of two kinds. Submerged leaves resembling stiff bladeless petioles, up to 50 cm long and 1 – 2 mm wide. Floating leaves (rarely absent) with petioles notably longer than the blades; the blades broadly elliptic, with commonly subcordate (occasionally cuneate or rounded) bases and rounded apices, up to 10 cm long by 6 cm wide, with 19 – 35 veins, usually coppery brown in colour in life. Transitional leaves, with very small blades, are sometimes found in spring. Sheaths free from petiole bases, stout, two-keeled, stiff and fibrous, up to 18 cm long, ultimately shredding into fibres.

Peduncles up to 12 cm long, emergent, thicker than the stems. Spike up to 3 cm long, emergent, thicker than the stems. Spike up to 3 cm long, compact and densely flowered, somewhat enlarged in fruit. Achene obovoid, 3 – 5 mm long, including a straight apical beak 0.5 – 1 mm long, khaki-brown when ripe; keeled or rounded dorsally when dry, the keel not conspicuous on fresh material. 2n = 52. Figure 33.

Widespread in temperate latitudes of the northern hemisphere, *P. natans* is found all over British Columbia. One of our commoner species of this genus, it is found in shallow to moderately deep waters (1 – 3 metres) of sheltered

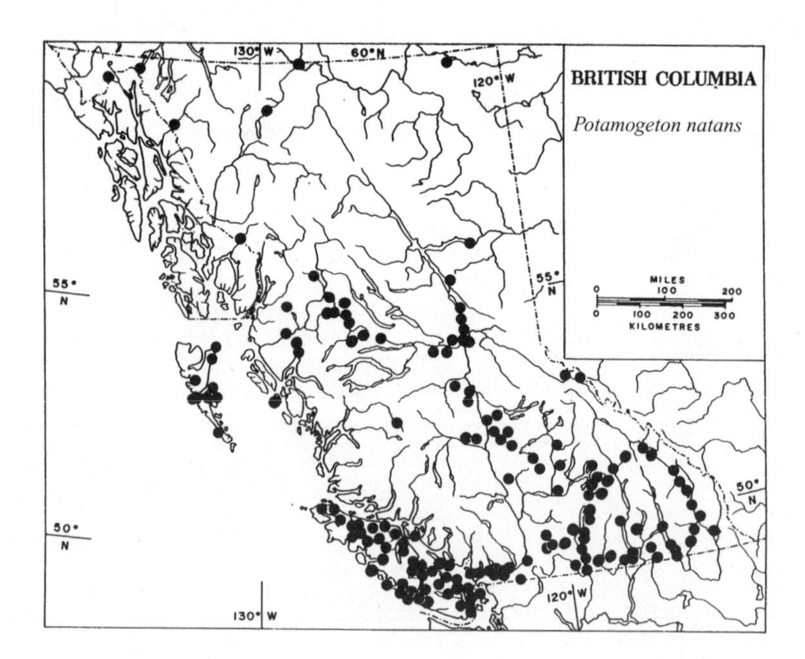

bays and ponds, often where the bottom is covered with organic detritus. Under some conditions, winter buds may be produced from the upper leaf axils. These buds bear leaves like abbreviated forms of submerged leaves. In mild areas, the entire shoot may survive over winter.

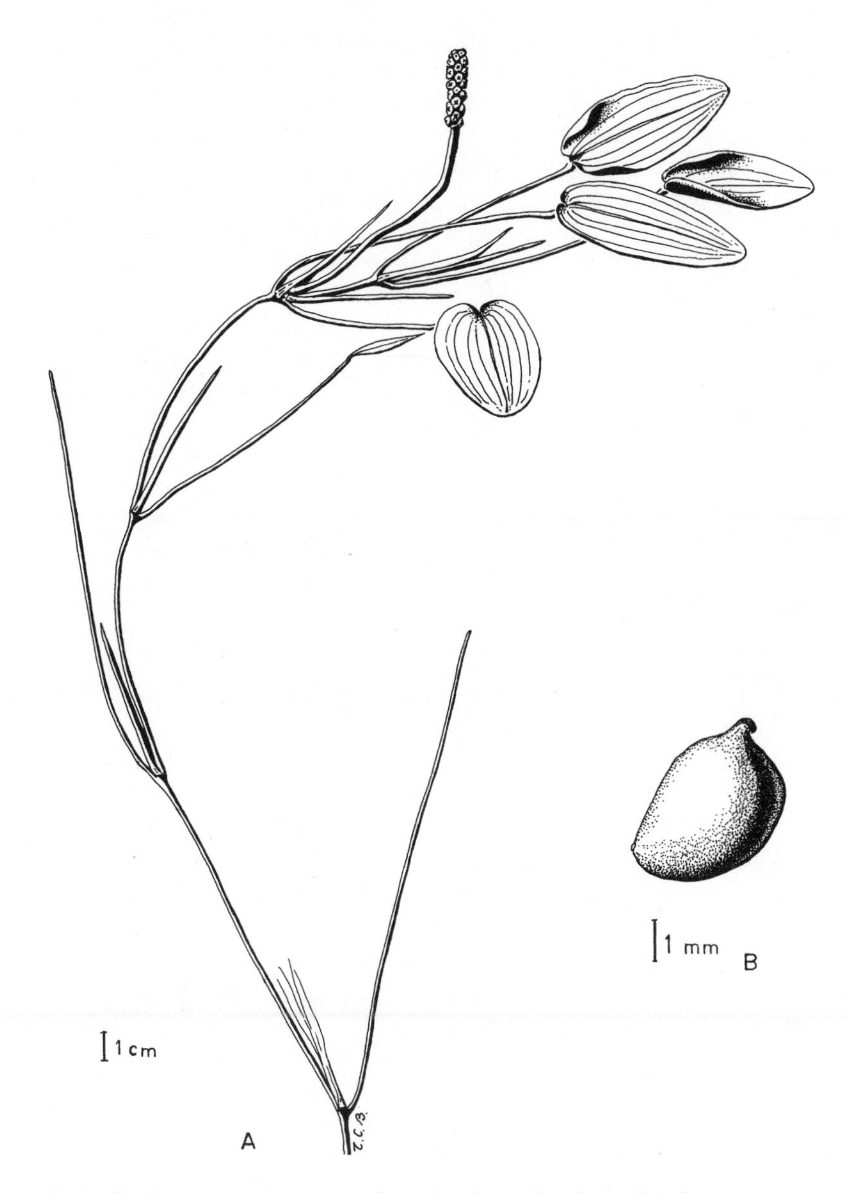

Figure 33. *Potamogeton natans*: A, flowering shoot; B, achene (dried).

Potamogeton nodosus Poiret **Long-leafed Pondweed**

Rhizomatous perennial with terete, unbranched or sparingly branched stems up to 2 metres long.

Leaves of two kinds, all petioled. Submerged leaves on petioles 2 – 10 cm long, with blades lanceolate to narrowly elliptic, 10 – 30 cm long by 1 – 3 cm wide, acuminate, with 7 – 15 longitudinal veins, thin and translucent. Floating leaves rather leathery, with elliptic blades 5 – 12 cm long by 1.5 – 4 cm wide, on petioles longer than the blades. Sheaths 3 – 8 cm long, free of the petioles and open to base on one side.

Peduncles thicker than the stems, up to 15 cm long. Spikes up to 6 cm long, densely flowered with up to 15 whorls of flowers. Achenes obovoid, obscurely keeled when fresh, but prominently keeled when dry, almost beakless, or with a very short beak less than 0.5 mm long. 2n = 52. Figure 34.

This species is widespread, mainly in warmer latitudes, throughout North America, South America, Eurasia and Africa. It is not common in British Columbia. It is found in shallow to deep clear fresh waters of lakes and ponds.

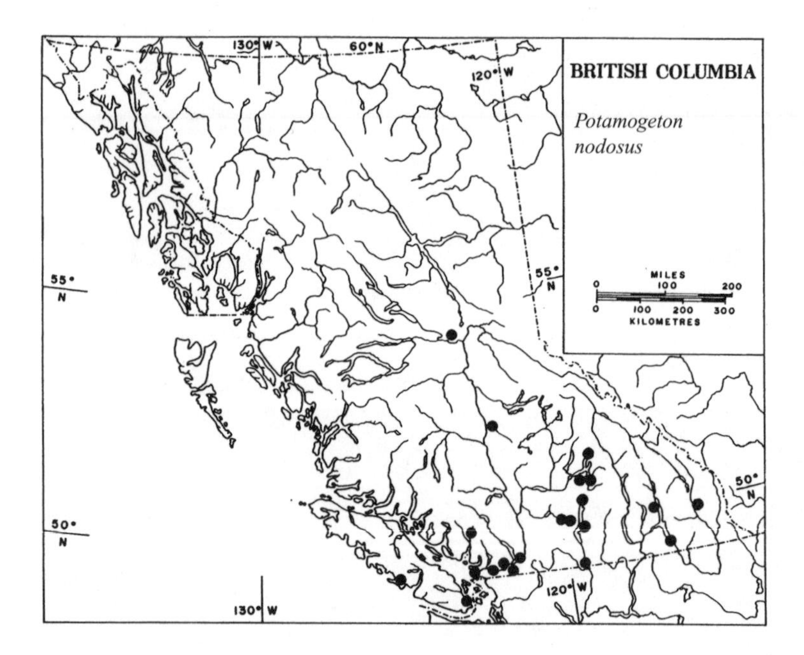

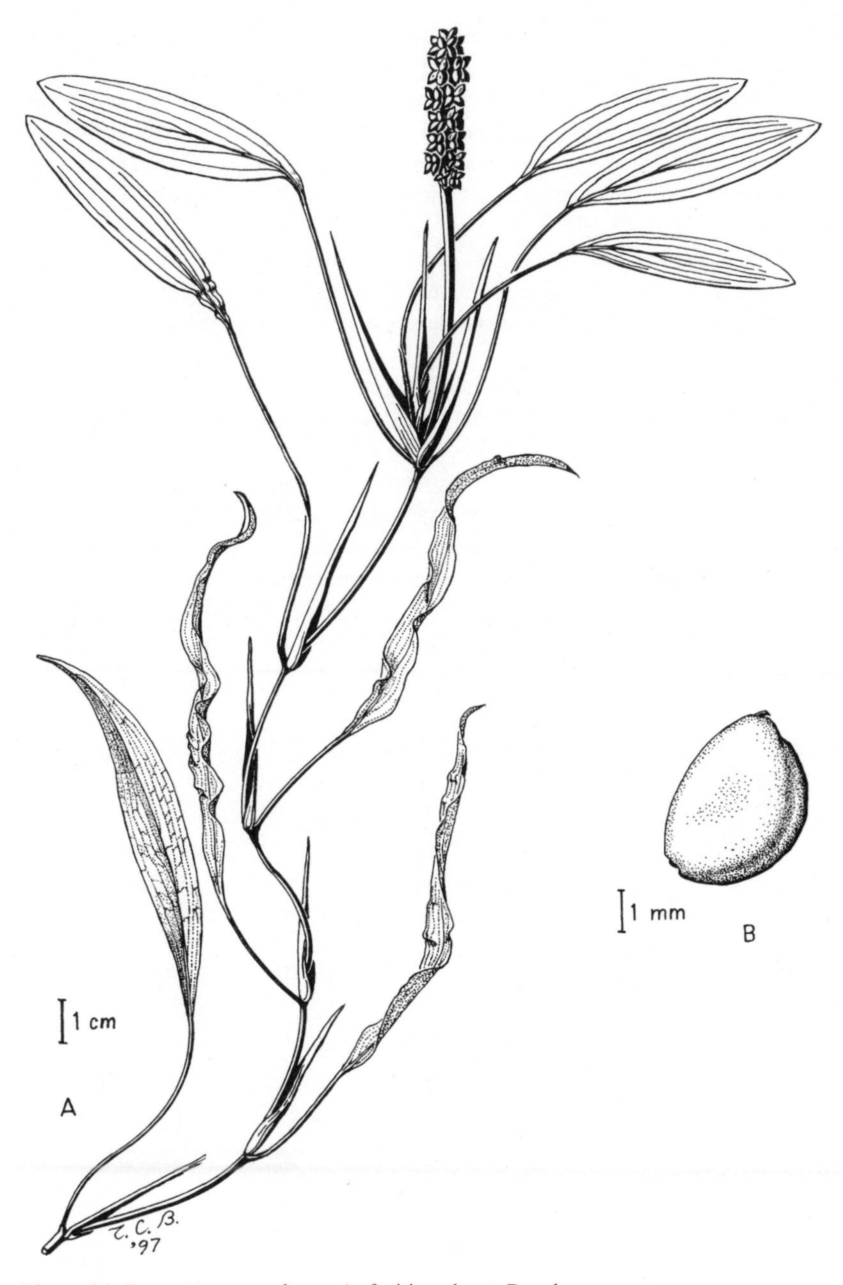

Figure 34. *Potamogeton nodosus*: A, fruiting shoot; B, achene.

Potamogeton oakesianus Robbins — Oakes' Pondweed

Like a small, delicate, more branched version of *P. natans*. Stem 0.5 – 1 mm thick, 0.5 – 1 metre long, with a branching pattern suggestive of *P. gramineus* var. *myriophyllus*. Submerged leaves linear, the upper ones dilated toward their apices, flat, thin, flaccid, 3 – 13 cm long by 0.5 – 2 mm wide, with one to three veins.

Floating leaf blades elliptic, acute at bases and apices, 7 – 25 veined, on long threadlike petioles. Sheaths membranous to somewhat fibrous, greenish and opaque, free from the petiole-bases, often becoming detached at their bases.

Spikes with three to eight whorls of flowers, up to 3.5 cm long by 7 – 9 mm thick in fruit. Achenes 2.5 – 3.7 mm long, with prominent dorsal keels. Figure 35.

A plant of eastern North America, this species has been found by Ceska and Ceska (1980) in a pond near Steelhead in the lower Fraser River valley, presumably as an introduction. The material collected there includes the material that was identified as *P. diversifollius* at the time of the first edition of this book. More recently, it has been found on Galiano Island.

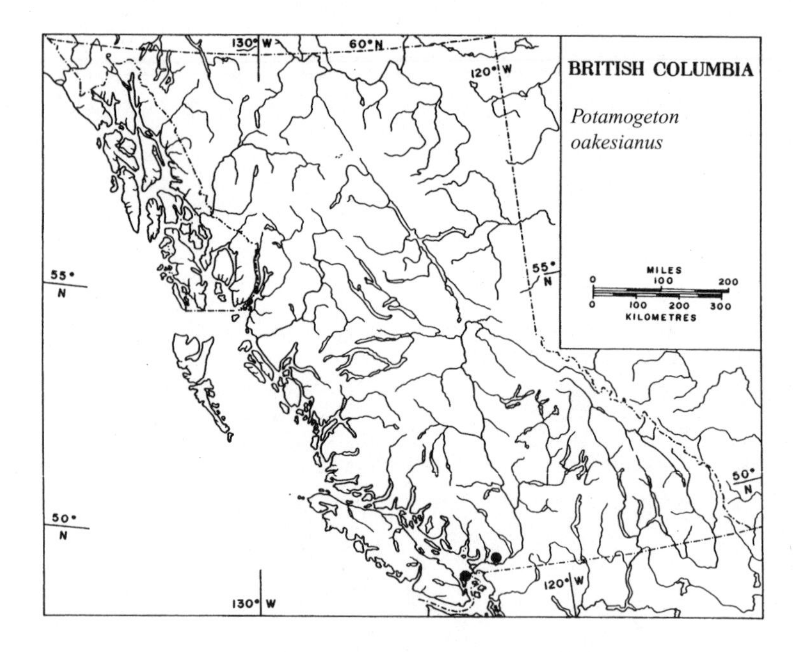

Figure 35. *Potamogeton oakesianus*.

Potamogeton obtusifolius Mertens and Koch **Blunt-leafed Pondweed**

Submerged plants 0.5 – 1 metre long, green or reddish, with terete or somewhat flat, slightly zigzag stems bearing paired swollen glands at the nodes. Perennating by densely leafy winter buds that terminate vegetative branches in autumn, and lie on the bottom of the water over winter to sprout in spring.

Leaves all alike, sessile, broadly linear, up to 10 cm long by 2 – 4 mm wide, rounded to sometimes acute at tips, translucent, green or reddish, with usually three veins (rarely more), the midvein bordered by two to four rows of air channels. Sheaths about 2 cm long, open to base on the side opposite the leaf base, sometimes rather flaring, eventually shredding into fibres from the tips.

Peduncles straight, short, up to 2 cm long. Spike short, 1 – 2 cm long, densely flowered. Achene 3 – 4 mm long, rounded or obscurely keeled dorsally, with a straight beak up to 0.7 mm long. 2n = 26. Figure 36.

Potamogeton obtusifolius is widespread in the northern hemisphere, and widely scattered but uncommon in British Columbia. It is found in shallow lakes and ponds, often with a bottom of organic detritus.

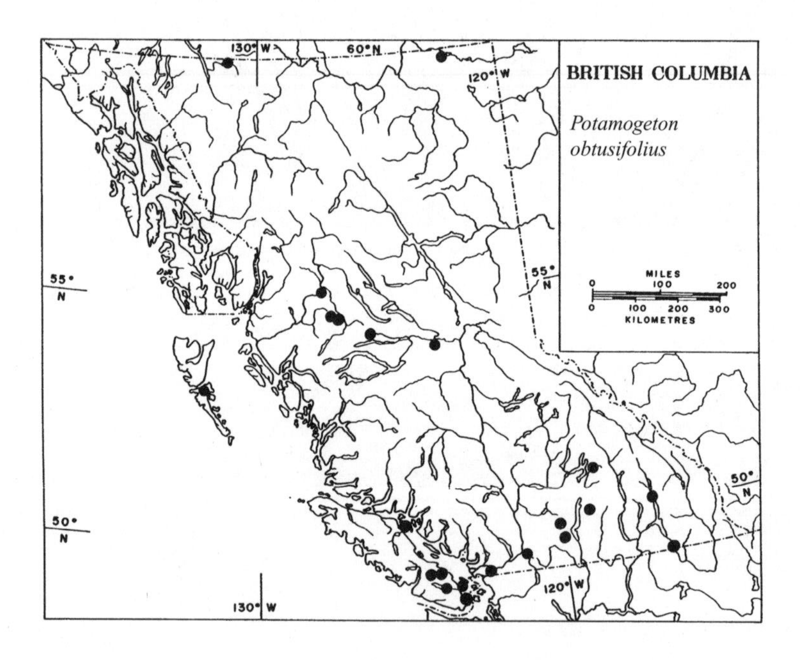

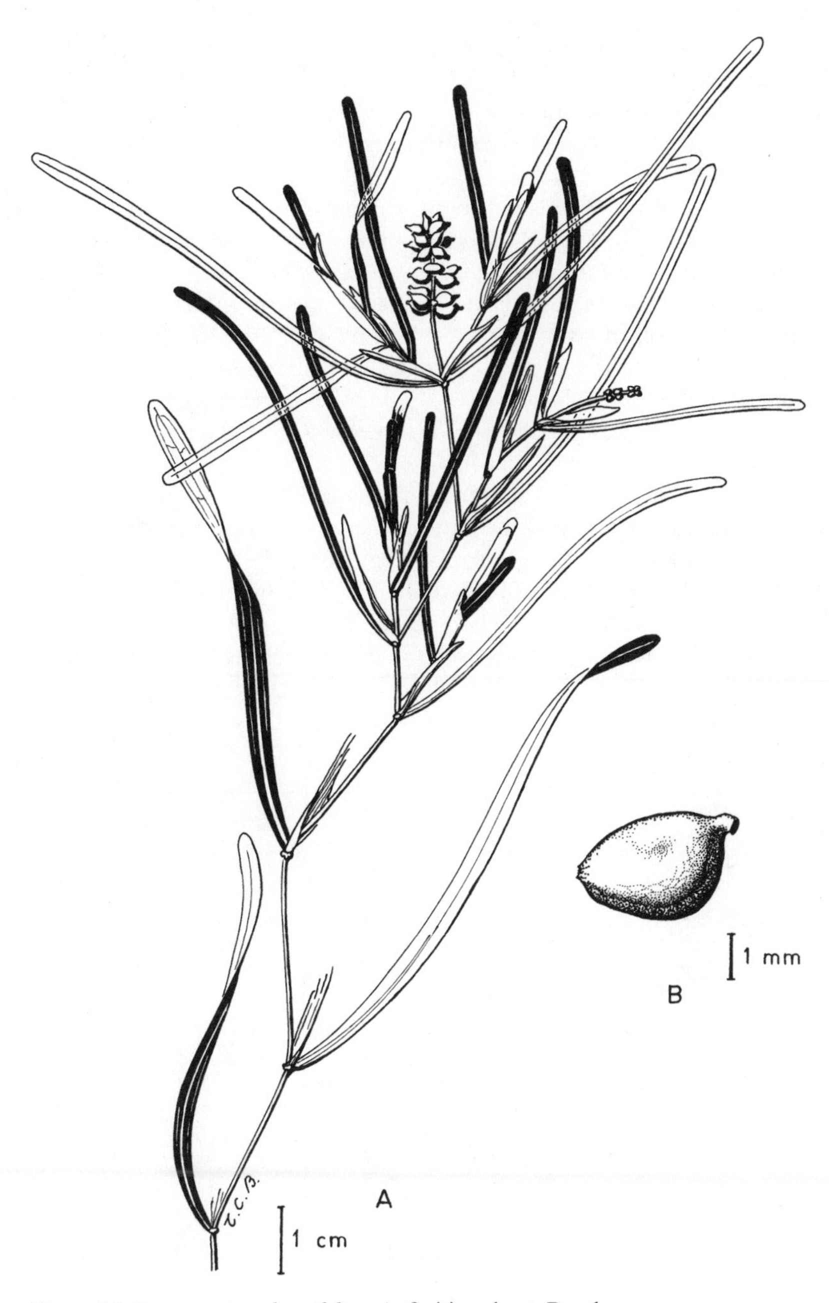

Figure 36. *Potamogeton obtusifolius*: A, fruiting shoot; B, achene.

Potamogeton pectinatus L. — Sago Pondweed

***P. marinum* L.?**
***Stuckenia pectinata* (L.) Borner**
***Coleogeton pectinatus* (L.) Les and Haynes**

Submerged plant with several stems arising from a buried, extensively branched, slender rhizome, which produces small overwintering tubers. Stems terete, filiform, frequently branching, with one, or sometimes two, branches in a leaf axil; the branches often having a rather featherlike aspect.

Leaves all alike, filiform, 2 – 12 cm long by up to 1 mm wide, tapering to acute tips, usually with one vein or, rarely, three. Sheath membranous, adnate for most of its length to the leaf base, 2 – 3 cm long, the margins overlapping but separate to the base on the side opposite the leaf; the ligule a tenth to a third the length of the whole sheath.

Peduncle filiform, lax, not raising the flower spike above the water surface, the spike at flowering time commonly submerged, or lying horizontally at or just beneath the water surface. Spike 1 – 2 cm long at flowering time, of two to six whorls of flowers, the lowest whorl often separate from the others. In fruit, the axis of the spike elongates up to 12 cm long, separating the whorls; the lowest whorls becoming remote from the others. Achene 3 – 4.5 mm long, with a beak up to 0.7 mm long, the achene reddish brown when ripe. 2n = 78. Figure 37.

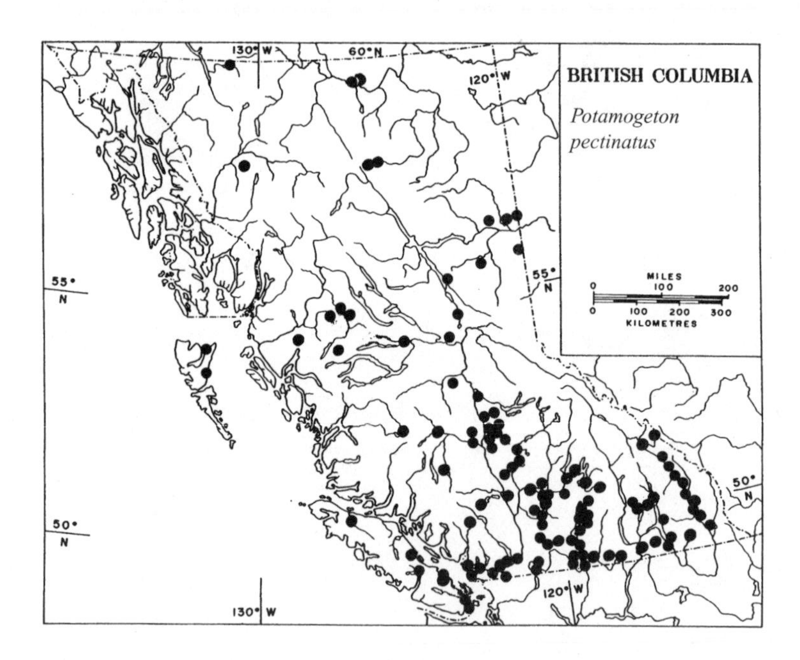

Potamogeton pectinatus is widespread in the northern hemisphere, and is common all over British Columbia, inhabiting still or slowly moving fresh water 0.5 – 2.5 metres deep. This species can tolerate a moderate degree of salinity or alkalinity, and is found in brackish marshes and very calcareous waters.

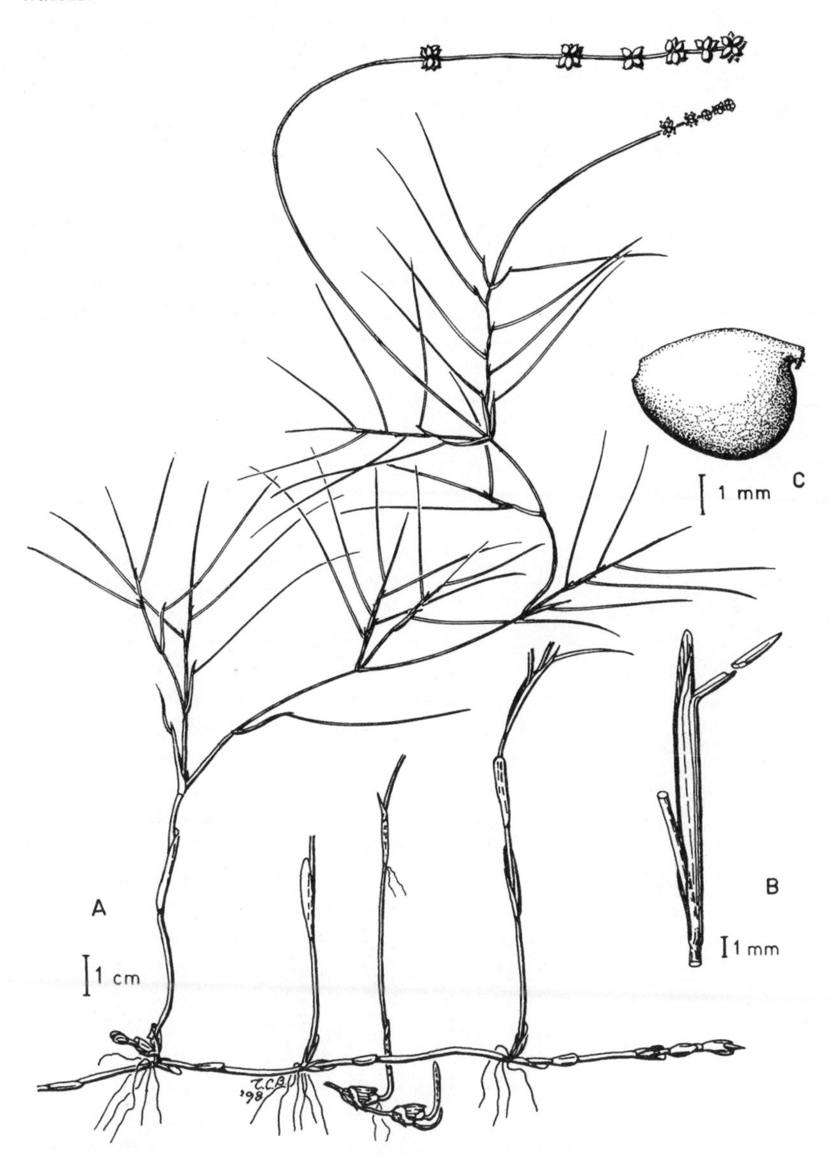

Figure 37. *Potamogeton pectinatus*: A, shallow water plant in flower and fruit, and tubers; B, sheath, leaf base and leaf apex; C, achene.

Potamogeton x *bottnicus* Hagstroem, the hybrid between *P. pectinatus* and *P. vaginatus*, has been found in Lake Windermere.

Note on the Nomenclature

Potamogeton marinum L. appears to be the same as *P. pectinatus* L., judging by specimens in the Linnaean Herbarium (Savage 1945). This name occurs on three sheets in that herbarium. Sheet No. 175.13, the only specimen collected by Linnaean himself (at Gotland), and labelled by him as *P. marinum*, bears a mature fruiting inflorescence, with the lowest two pairs of flowers remote from the others; and at least a few of the achenes appear distinctly beaked. This specimen is clearly *P. pectinatus* as we now understand this species. Sheets Nos. 175.12 and 175.14 were received by Linnaeus from Philip Miller, and labelled by Miller as *P. pectinatus*. They lack fruiting inflorescences, and were re-identified by Linnaeus as *P. marinum*. Both names occur in Linnaeus' *Species Plantarum* (1753), but no specimen remains in his herbarium that he identified as *P. pectinatum*.

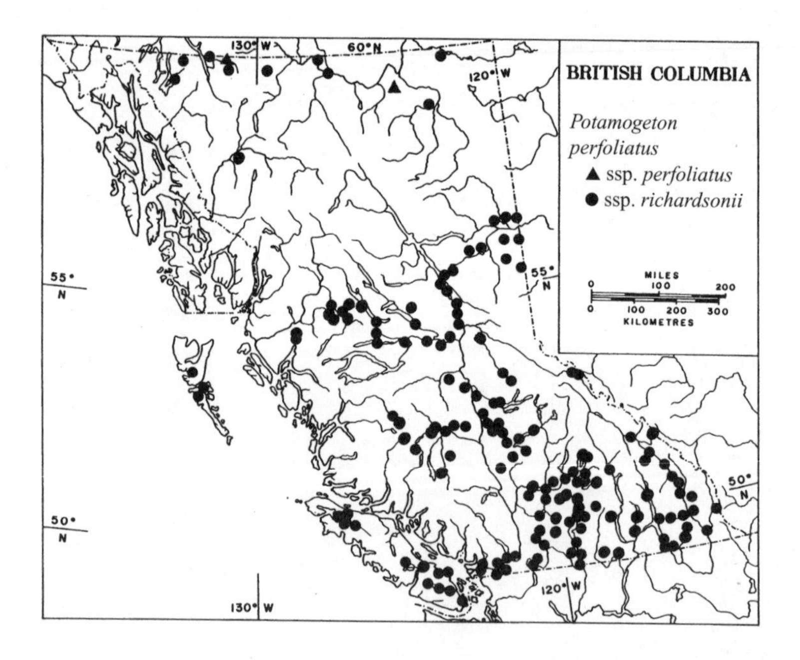

Potamogeton perfoliatus L. — Red-head Pondweed

***P. perfoliatus* var. *richardsonii* Bennett, *P. richardsonii* (Bennett) Rydberg**
= *P. perfoliatus* ssp. *richardsonii* (Bennett) Hulten

Sparsely branched, immersed plant with long terete stems bearing leaves at frequent and regular intervals and spreading by buried creeping rhizomes.

Leaves all submerged and alike, sessile, broadly ovate to lanceolate, 2 – 13 cm long by 1 – 3 cm wide, clasping the stem by their cordate bases, with seven to many longitudinal veins and fine cross-veins, the margins with minute one-celled teeth. Sheaths variably membranous with few to many incorporated fibres, tending to disintegrate early.

Peduncles commonly short and stout, moderately dilated above, often becoming recurved; commonly little longer than the spike, but sometimes very elongated. Spike short, dense, of 4 – 12 whorls of flowers, 2 – 4 cm long, becoming 1 cm thick in fruit. Achene 2.5 – 4 mm long, including the 0.5 – 1 mm long beak. Usually without a keel. 2n = 52. Figures 38 and 39.

Key to Subspecies

1a. Stem and peduncle relatively slender. Leaves broadly ovate, rounded at apex, often straight, occasionally recurved. Sheath membranous, with few short fibres, soon disappearing. Achene without a cavity in the embryo loop..ssp. *perfoliatus*

1b. Stem stouter. Peduncle thickened distally. Leaves more narrowly ovate to lanceolate, acute at apex, often recurved and with undulate margins. Sheath with many strong fibres that persist as the membranous parts decay. Achene with a cavity in the embryo loop. ....ssp. *richardsonii*

Subspecies *richardsonii* (Bennett) Hulten, the common subspecies in North America, is abundant throughout British Columbia. Probably our most abundant species of pondweed, it is found in shallow to deep clear lake waters, often where slightly calcareous, or in slow-flowing rivers. This species shows much variation among individuals in such characters as the spacing, size and form of the leaves, and the length of the peduncle. This variation is little understood, but the basis appears to be partly genetic, and partly a response to conditions such as the depth of water and the advance of the season's growth (Spence and Dale 1978).

Subspecies *perfoliatus*, the common subspecies across Eurasia, is also found along the Atlantic coast of North America and inland to Lake Erie, as well as in central Alaska (Hulten 1941), Yukon and extreme northern British Columbia. In this province, it has been found at Swan Lake in the Cassiar Range (59° 54' N, 131° 21' W), northwest of Fort Nelson.

Hulten subsequently (1968) modified his earlier statement, and on the basis of the achene character given in the key, assigned all northwestern North American material to ssp. *richardsonii*. Unfortunately, most of our material of ssp. *perfoliatus* from that area lacks the achenes that would clinch its determi-

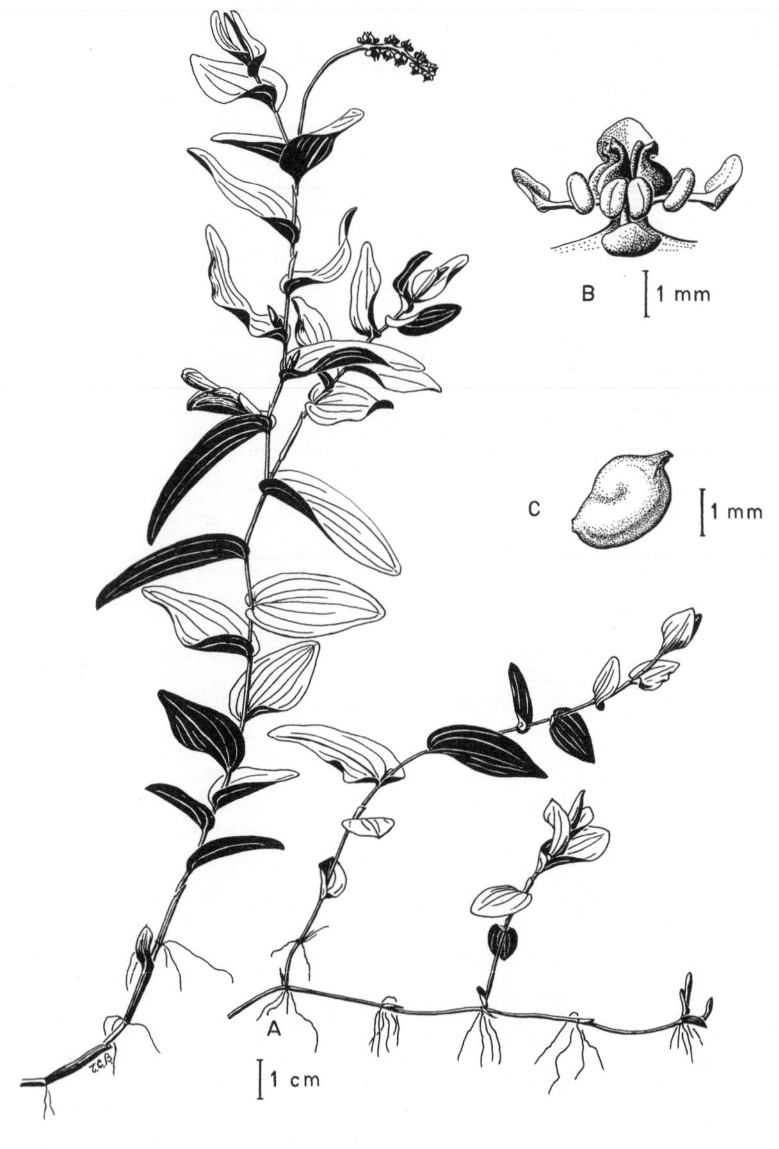

Figure 38. *Potamogeton perfoliatus* ssp. *perfoliatus*: A, shallow water plants; B, flower; C, achene.

nation, so its identity is based on its visible vegetative characters. More collections of this plant from that area, with fruiting material, are desirable.

Where ssp. *perfoliatus* and ssp. *richardsonii* grow in proximity to each other, as in Swan Lake, intermediate forms are common.

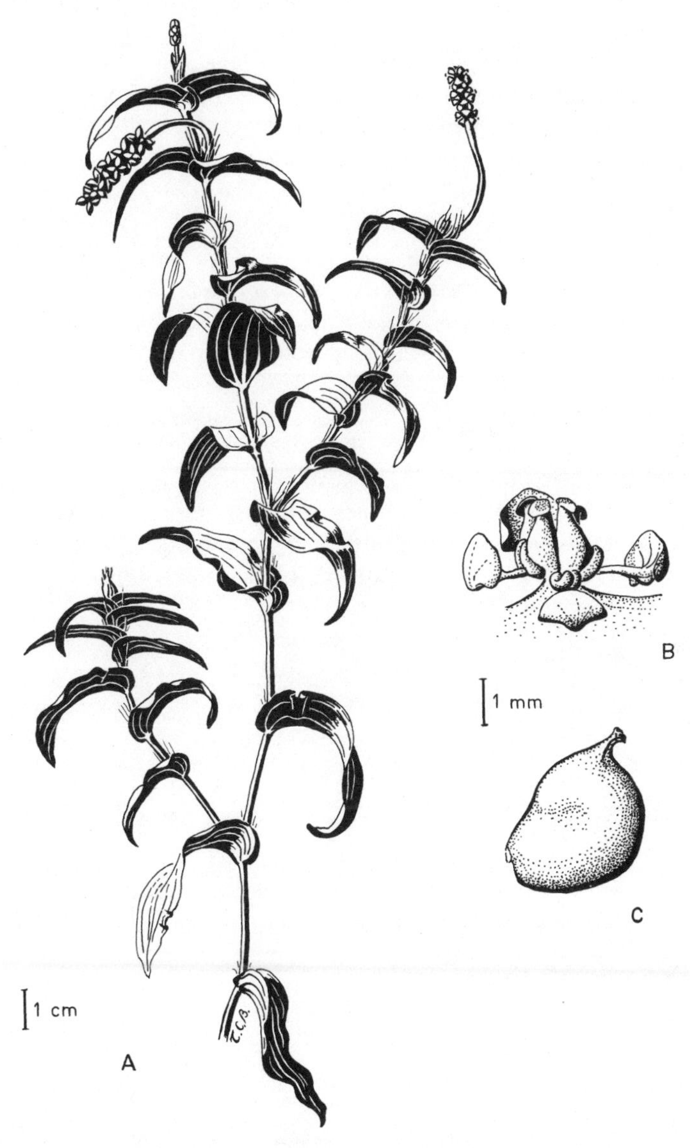

Figure 39. *Potamogeton perfoliatus* ssp. *richardsonii*: A, upper part of plant, flowering and fruiting; B, flower; C, achene.

Potamogeton praelongus Wulfen **White-stem Pondweed**

The largest of our pondweeds. Immersed rhizomatous perennial with a stout, brown-spotted white rhizome 2 – 6 mm thick containing numerous air passages. Erect stems 2 – 6 metres long and 2 – 3 mm thick, with stout, shiny, often zigzag upper stems producing frequent branches toward the top, though sparingly branched below.

Leaves all alike in general form, sessile, shiny, 7 – 30 cm long by 15 – 25 mm wide, commonly clasping the stem by their cordate or subcordate bases, widest below the middle, and tapering gradually to rounded, hooded tips, with five or more stout longitudinal veins and finer intermediate veins and cross-veins; the margins entire, or minutely serrulate on the hooded apex. Leaves on the short axillary branches are generally smaller than those on the upper main stem, and deeper leaves are generally shorter, and less cordate, or even cuneate at their bases. Sheaths free from the leaf bases, stiff, whitish, 3 – 10 cm long, persistent or eventually disintegrating.

Peduncles emergent, stout, 10 – 50 cm long, somewhat thickened upward. Spikes up to 5 cm long with 6 – 12 close-set whorls of flowers. Achenes 4 – 8 mm long, including the very stout 0.5 – 1 mm long beak, and distinctly keeled dorsally. 2n = 52. Figure 40.

Circumboreal in distribution, south in North America into much of the United States, *P. praelongus* occurs widely across British Columbia. Typically found in deep clear lakes, it may grow in water up to 6 metres deep. It is capable of a phenomenal rate of growth in spring – shoots having been found up to

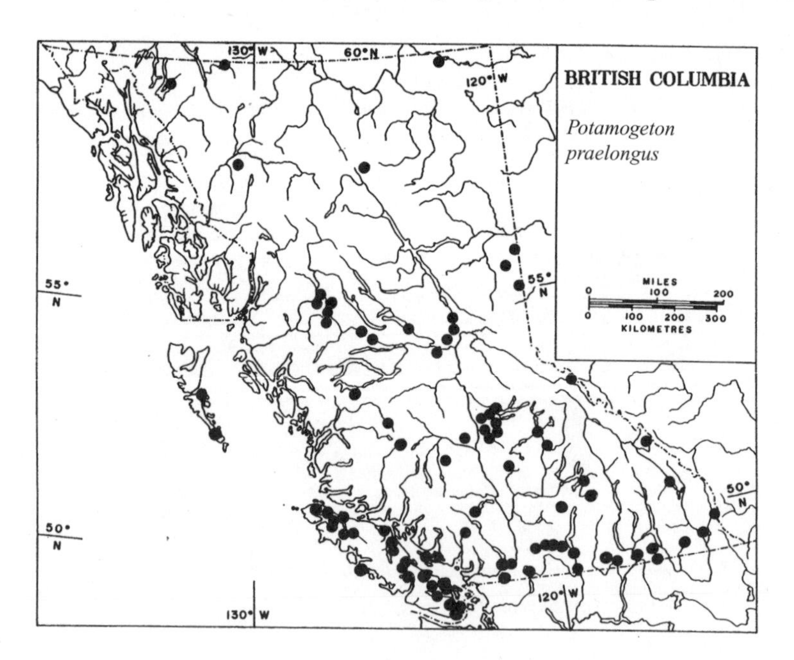

5 metres long by the beginning of May in the Victoria area. Late in the growing season, large winter buds may be formed in the upper leaf axils, for vegetative perennation. These winter buds are abbreviated branches, with short internodes, and crowded, diverging, short leaves emerging from among conspicuous white sheaths.

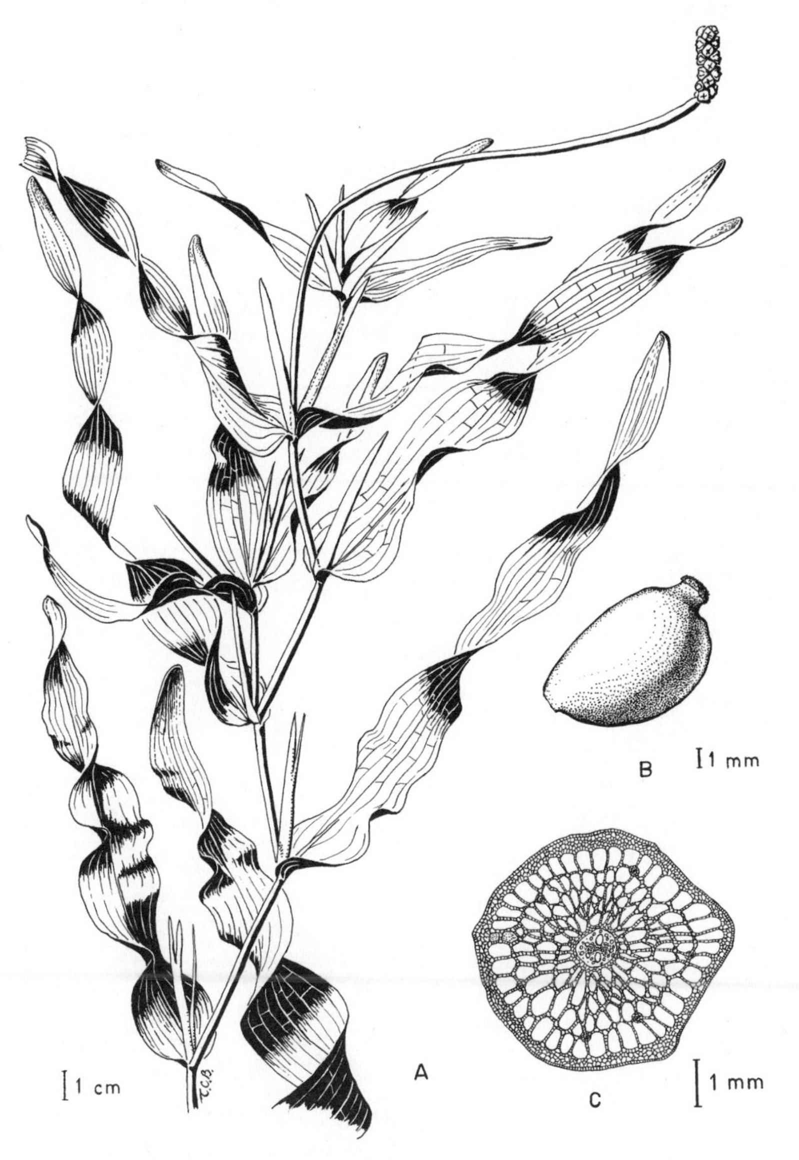

Figure 40. *Potamogeton praelongus*: A, upper part of plant; B, achene; C, cross-section of rhizome, showing spongy cortex (lacunate tissue) traversed by air passages (lacunae) separated by thin partitions, and the slender central vascular strand.

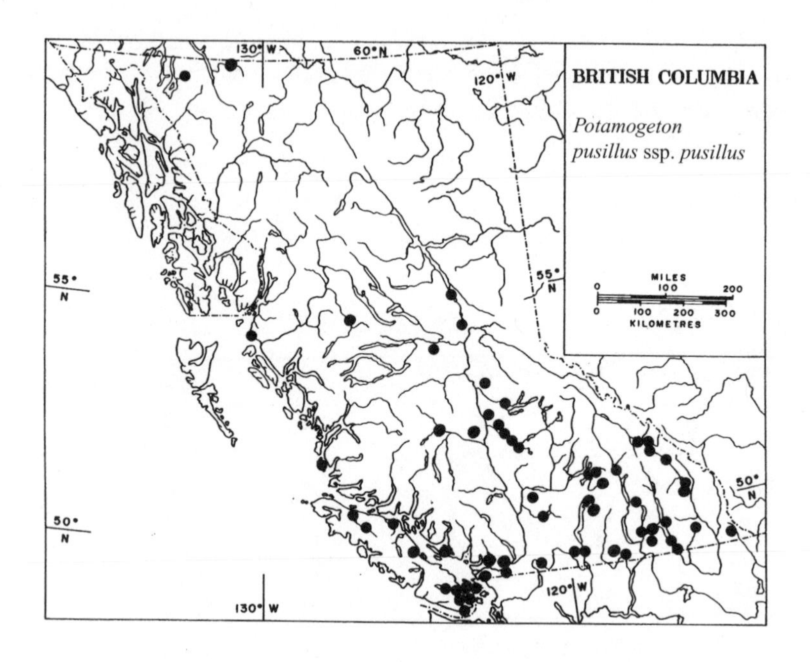
BRITISH COLUMBIA
Potamogeton pusillus ssp. pusillus
MILES
0 100 200
0 100 200 300
KILOMETRES
130° W
60° N
120° W
55° N
50° N
120° W
130° W

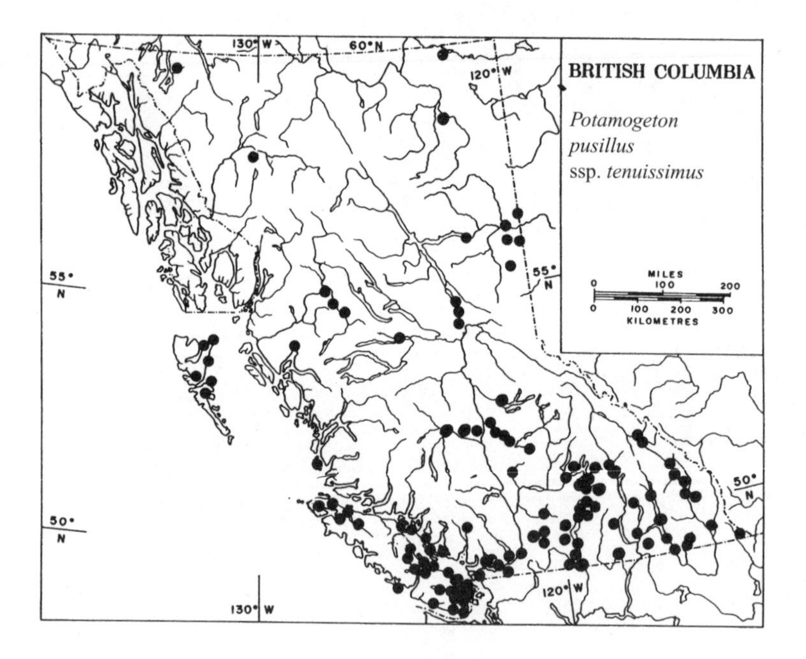
BRITISH COLUMBIA
Potamogeton pusillus ssp. tenuissimus
MILES
0 100 200
0 100 200 300
KILOMETRES
130° W
60° N
120° W
55° N
50° N
120° W
130° W

Potamogeton pusillus L. Small-leafed Pondweed

***P. panormitanus* Bivona-Bernardi**
***P. berchtoldii* Fieber, including var. *tenuissimus* (Mertens and Koch) Fernard and var. *lacunatus* (Hagstrom) Fernald = *P. pusillus* L. ssp. *tenuissimus* (Mertens and Koch) Hayes and Hellquist (var. *tenuissimus* Mertens and Koch)**

Fine structured, usually profusely branched, submerged plant, up to a metre long, with slender, terete stems and often widely diverging branches, normally, but not invariably, bearing pairs of yellowish glands at the nodes. Generally non-rhizomatous, and perennating by small winter buds formed in the leaf axils or on the ends of lateral branches late in the growing season.

Leaves all alike, submerged, sessile, linear 2 – 7 cm long by 0.5 – 2 mm wide, normally three-veined, acute to obtuse or rounded at the tips. Sheaths free from the leaf bases, delicate, membranous, translucent pale greenish, usually shed or disintegrating long before the leaves do; often 1 cm or less, but occasionally up to 3 cm long.

Peduncles slender, 2 – 3 (– 5) cm long, often strongly curved near base. Spike 3 – 15 mm long, with one to four whorls of flowers, usually raised above the water surface briefly at flowering time. Achenes 1.5 – 3 mm long, rounded and without dorsal keels, with a short beak up to 0.5 mm long; ripening under water. 2n = 26. Figures 41 and 42.

Key to Subspecies

1a. Leaf with no air spaces (lacunae) or at most only one row on either side of the midvein. Sheath margins connate when young, later ruptured by growth of axillary branches. Spike up to 15 mm long with two to four whorls of flowers. Fruit widest above the middle. ....ssp. *pusillus*

1b. Leaf with one to five rows of lacunae each side of the midvein. Sheath margins separate from the first, though often overlapping. Spike up to 7.5 mm long with one to three whorls of flowers. Fruit widest at or below mid-length. ....ssp. *tenuissimus*

Distinction between these two entities, either as varieties or species, has long been a source of confusion. No single character is 100 per cent reliable in separating them: a consensus of all characters is preferable. The sheath character is commonly treated as decisive, but does not correlate perfectly with other characters. Care must be used in handling the sheaths, which are easily and soon split; and only the youngest ones should be used. The treatment of this complex in Haynes 1974 and Haynes and Hellquist 1996, uniting *P. berchtoldii* (including its var. *tenuissimus* and var. *lacunatus*) with *P. pusillus* as ssp. *tenuissimus*, is regarded as the most realistic, and is followed here.

Potamogeton pusillus is widespread in the northern hemisphere, and ranges all over North America. In British Columbia, it is found scattered all across the province; subspecies *tenuissimus* being the commoner subspecies. It is

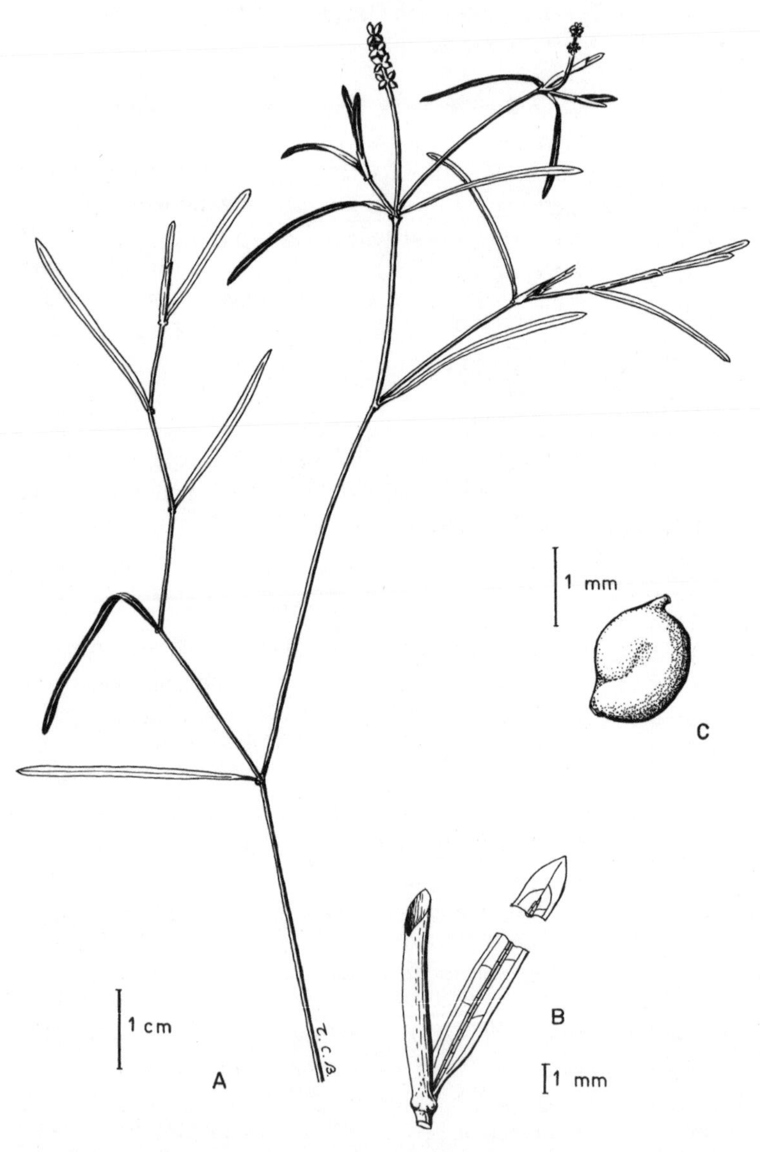

Figure 41. *Potamogeton pusillus* ssp. *pusillus*: A, fertile branch; B, detail of node, with nodal glands, sheath, and leaf base and apex; C, achene.

typically found in shallow fresh ponds and small lakes in water up to a metre deep, often in rather acidic water. The leaves, while alike in form, may vary widely in size with the season, even on the same plant, and may also vary with the habitat.

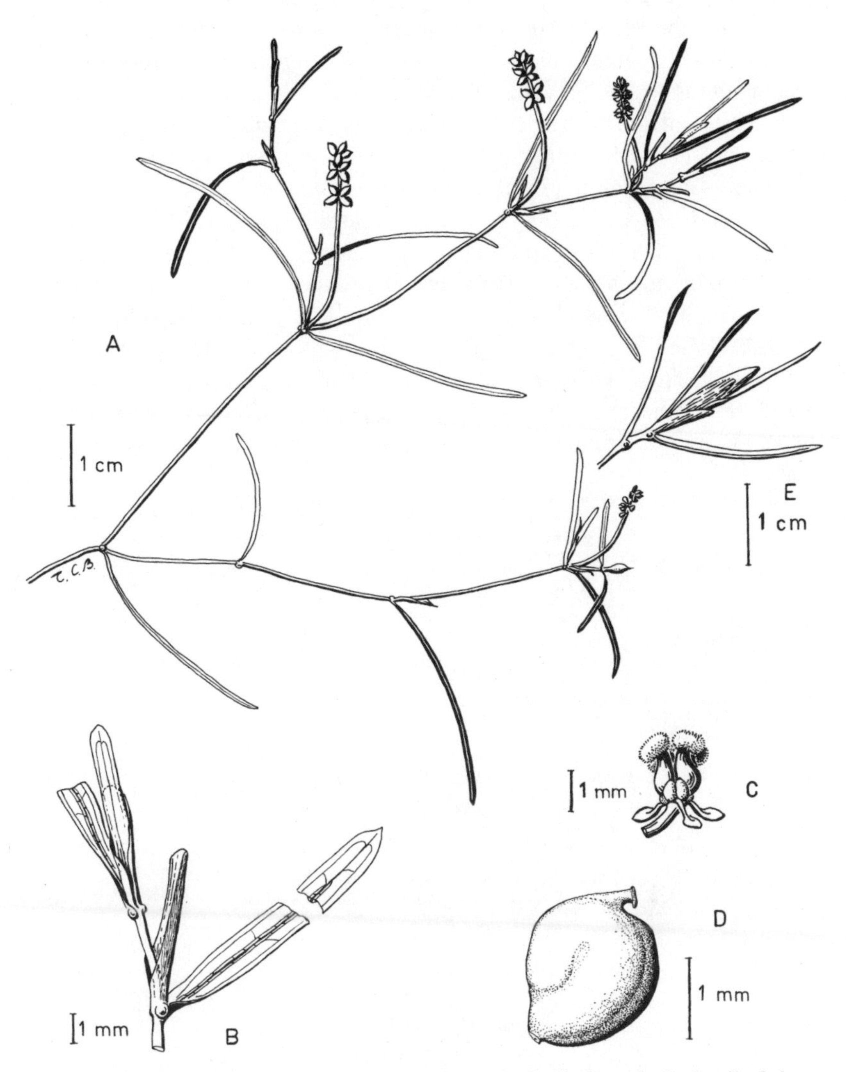

Figure 42. *Potamogeton pusillus* ssp. *tenuissimus*: A, fertile branch; B, detail of shoot tip, nodal glands, sheath and leaf; C, flower; D, achene; E, winter bud.

Potamogeton robbinsii Oakes **Robbins' Pondweed**

Completely immersed plant, with a stem creeping on the bottom of the water, rooting at the nodes, moderately compressed, 0.5 – 3 metres long. Leafy shoots prostrate and creeping; the flowering shoots ascending toward the water surface.

Leaves crowded in one plane and obviously two-ranked, stiff and widely diverging from the stem, linear-lanceolate, up to 12 cm long by 6 mm wide, dark olive to brownish green, and often covered by colonies of diatoms and other algae, with many parallel veins, and white, finely serrulate margins; the subcordate base of the blade arising from about the mid-point of the adnate sheath. Sheath 10 – 15 mm long, white, fibrous, shredding from the tip into its constituent fibres; commonly covering the stem between the leaves.

Flowering shoots ascending to near the water surface, with longer internodes and shorter leaves than on vegetative shoots; branching and producing several flower spikes on flat, stiff, slender peduncles, each spike with about four whorls of flowers. This species flowers late (July to September), the whorls becoming shortly separated at the fruiting stage. Mature achenes 3 – 4 mm long, blackish, with a distinct dorsal keel and a curved beak up to 1 mm long. Mature achenes are seldom produced, and reproduction is generally vegetative by overwintering shoot tips or adjacent, coarse winter buds. 2n = 52. Figure 43.

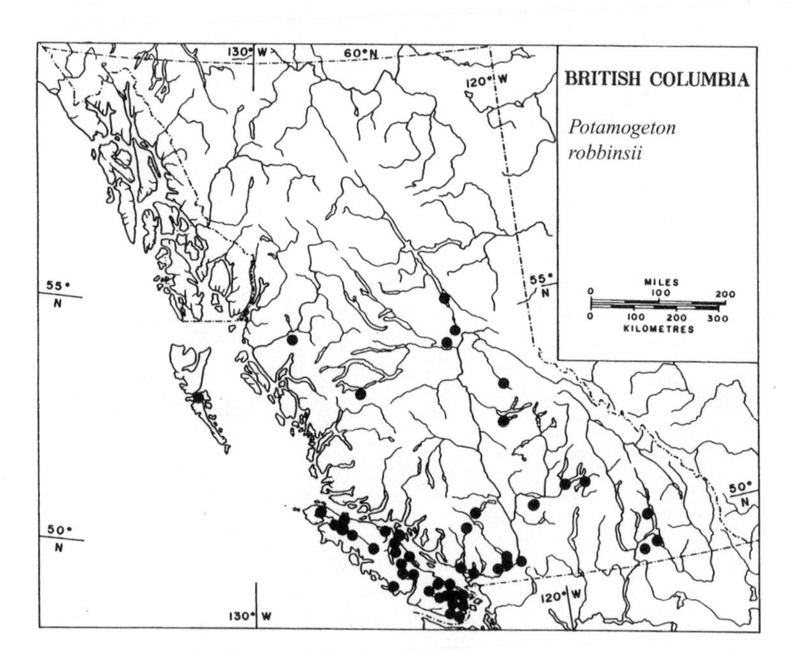

This species is widespread, though not continuous, across Canada and the northern United States. In British Columbia it has been found all across the province south of about 55°N latitude.

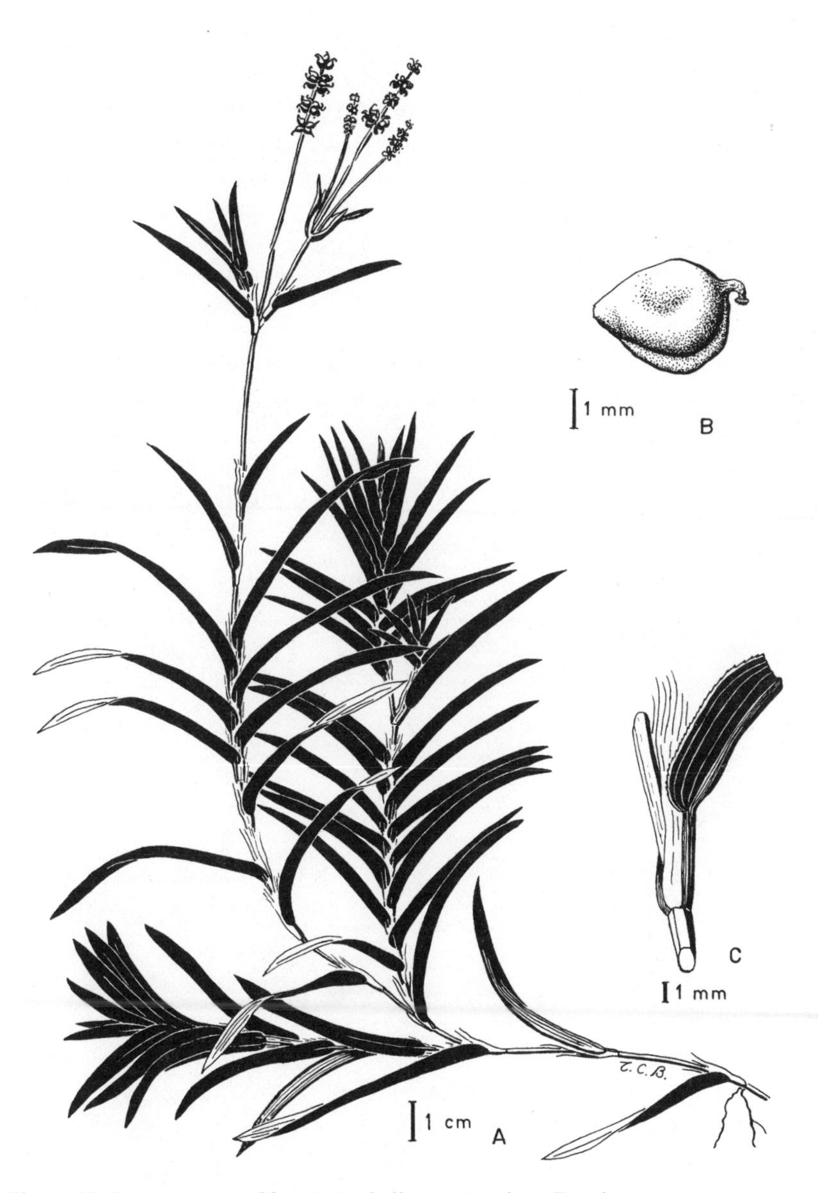

Figure 43. *Potamogeton robbinsii*: A, shallow water plant; B, achene; C, sheath and leaf base.

Potamogeton strictifolius Bennett Stiff-leafed Pondweed

Somewhat resembling *P. friesii*, but rather sparingly branched, with relatively short internodes and crowded foliage. Stems somewhat flat, but rather less so than in *P. friesii*, with paired glands at the nodes.

Leaves all alike, linear, 3 – 5 cm long by 0.5 – 2.5 mm wide, stiff, acute at apices, with three (or occasionally five) longitudinal veins and few or no cross-veins, olive green to brownish. Midvein prominent, not bordered by lacunae, or by only one inconspicuous row each side. Margins often rolled under, appearing thickened. Sheaths white, very fibrous, with connate margins below when young, but becoming frayed into fibres with age, 1 – 2 cm long, often longer than the internodes and largely concealing the stem.

Peduncle filiform, thickened above, 1 – 9 cm long. Spike 1 – 1.5 cm long, open, with three or four whorls or pairs of flowers. Achene 2 – 3 mm long, beaked. 2n = 52. Figure 44.

Reproducing vegetatively by slender winter buds terminating sterile branches, and covered by appressed sheaths and corrugated-based short blunt leaves.

Known from Saskatchewan to the northeastern United States. This species is rare in British Columbia, having been found so far only in Windermere and Columbia lakes, and at Kawkawa Lake near Hope (Ceska and Ceska 1980). Fruit has, so far, not been found in British Columbia.

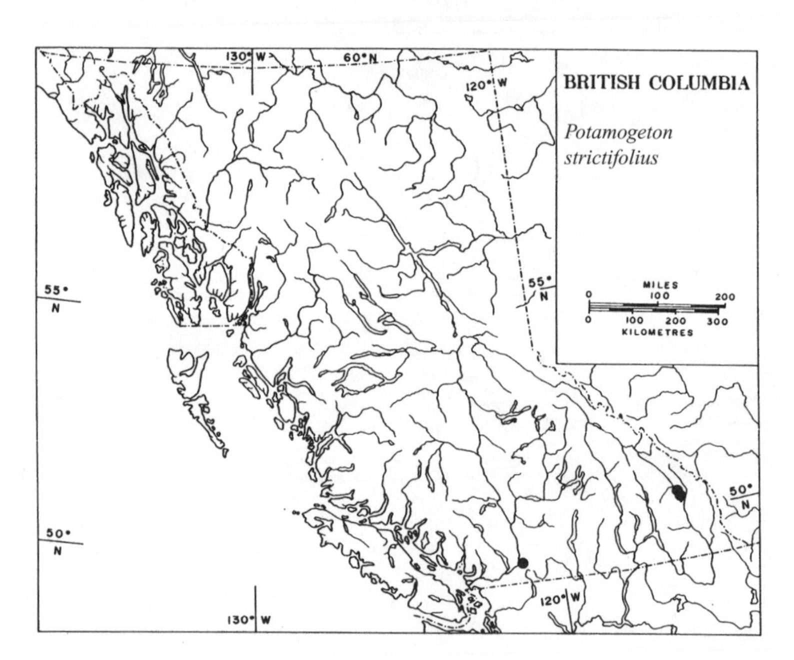

Figure 44. *Potamogeton strictifolius*: plant with flowering shoot, and vegetative branches terminating in winter buds.

Potamogeton vaginatus Turczaninow **Sheathed Pondweed**

***P. marinum* L.?**
***Stuckenia vaginata* (Turcz.) Borner**
***Coleogeton vaginatus* (Turcz.) Les and Hayes**

Immersed rhizomatous perennial, resembling *P. pectinatus*, but generally coarser, the stem terete, stout below but more slender above, freely branching, the lower stem commonly giving off two (sometimes three or more) branches at a node.

Leaves all alike, sessile, adnate to the sheaths, linear, up to 10 cm or more long by 1 – 2 mm wide, rather thick, with one to three veins, and rounded to obtuse at tips, often dark brownish in colour. Sheaths adnate for most of their lengths to the leaf bases, distended, conspicuously so on the lower main stem, where they may enclose the bases of multiple branches, firm textured, often greenish with whitish margins that are not connate though commonly overlapping; the ligules are short, a tenth to a quarter the length of the whole sheath.

Peduncle up to 12 cm long, slender and flexible. Spike floating at flowering time, with five to nine closely and evenly spaced whorls of flowers. Achenes obovoid, 3 – 4 mm long, obscurely keeled or rounded dorsally, and beakless, the stigma sessile. 2n = 78. Figure 45.

P. vaginatus is circumboreal in distribution, but rather scattered and less commonly seen than *P. pectinatus* or *P. filiformis* in British Columbia. It inhabits cold lakes and slow rivers, especially where the water is somewhat calcareous, in water 1.5 – 3 metres deep, generally deeper than that in which *P. pectinatus* is found.

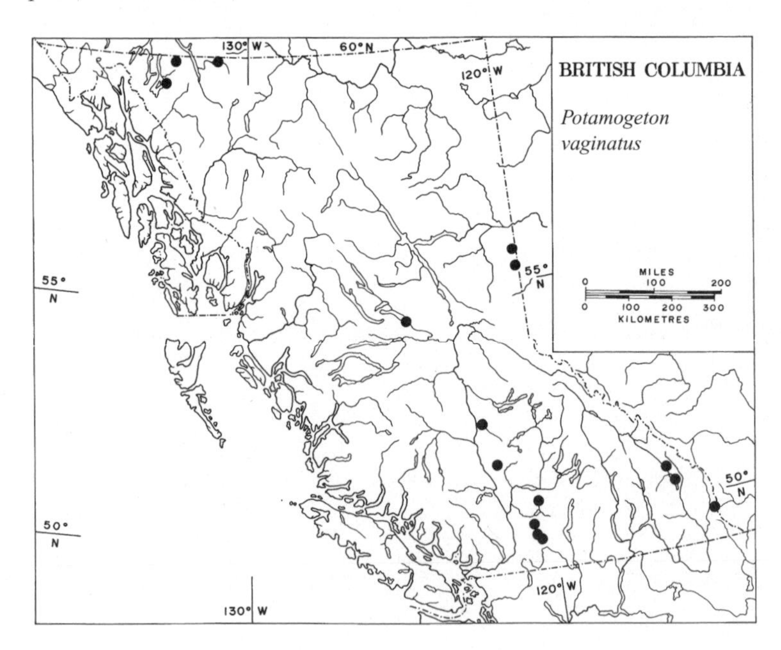

Potamogeton vaginatus is connected to the related *P. filiformis* through a number of intermediate forms. This putative hybrid population is known as *P.* x *fennicus* Hagstrom.

Note on the Nomenclature

The name *P. marinum* L. for *P. vaginatus* is suggested by Linnaeus' (1753) diagnosis: "Leaves linear, alternate, with the lower ones distinctly sheathing." But it is not suggested by the specimens in the Linnaean Herbarium. See also the note under *P. pectinatus*.

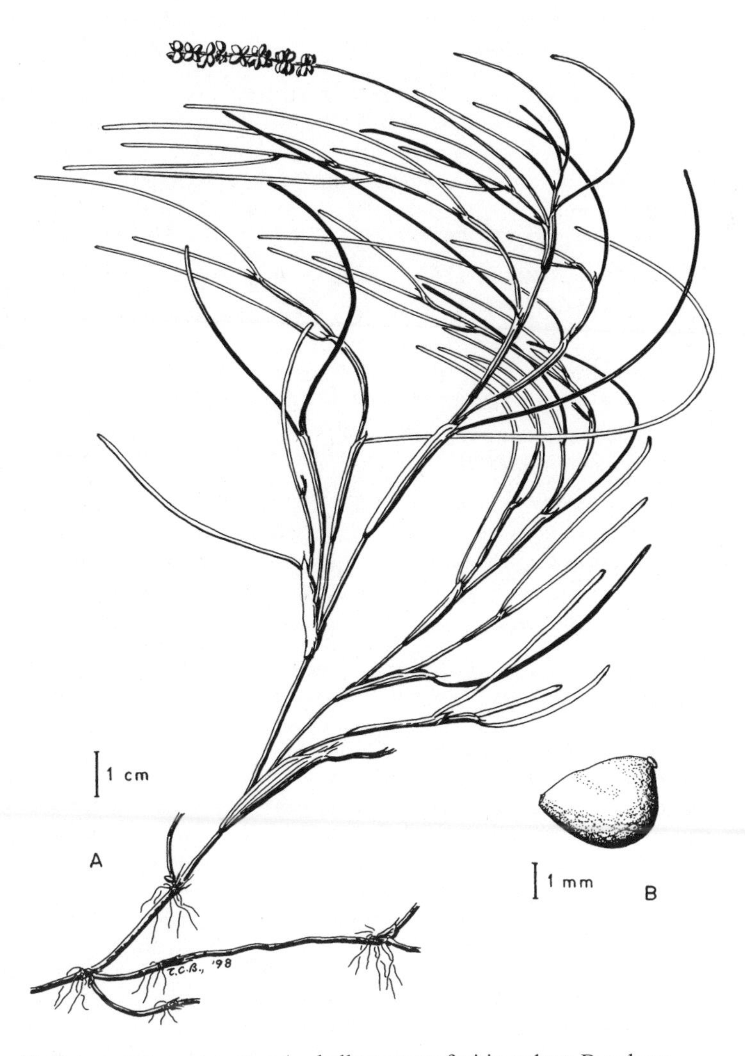

Figure 45. *Potamogeton vaginatus*: A, shallow-water fruiting plant; B, achene.

Potamogeton zosteriformis Fernald Eel-grass Pondweed

***P. zosterifolius* Schumacher ssp. *zosterformis* (Fernald) Hulten**
Not *P. compressus* L. (= *P. zosterifolius* Schumacher, in the strict sense)

Immersed perennial, sometimes rhizomatous, with flat, wing-edged, freely branching zigzag stems up to 2 metres or more long, and 2 – 3 mm wide, but constricted somewhat at the nodes, and often with paired small greenish nodal glands. Perennating by winter buds 4 – 7 cm long.

Leaves all alike, sessile, linear, up to 20 cm long by 3 – 5 mm wide, scarcely narrowed at bases; the early leaves rounded at their apices, later ones mucronate to acuminate with five longitudinal veins, scattered cross-veins, and numerous fine longitudinal veinlike interneural strands of sclerenchyma with free ends. Sheaths free from the leaf bases, open to base on the side opposite the leaf base, firm and fibrous, white eventually shredding into fibres, 2 – 6 cm long.

Peduncle flat, up to 10 cm long. Spike 2 – 2.5 cm long, closely flowered. The flower has only one carpel, one of the upper two. Fresh achene angularly broadly elliptic to suborbicular, 4 – 6 mm long, with a pronounced undulate and irregularly crested dorsal keel, and a stout, slightly curved stylar beak 0.6 – 1 mm long. Lateral keel-like wrinkles may appear on dried achenes. 2n = 52. Figure 46.

Potamogeton zosteriformis is transcontinental, widespread in Canada and in the northern United States. It is not uncommon in lakes in the interior of British Columbia, in water 1 – 2.5 metres deep, but is less common further north and on the coast. (Map on p. 148.)

The closely related *P. compressus* L. of Europe differs in having, on the average, more slender stems, less constricted at the nodes, more fibrous sheaths, smaller and more narrowly elliptic achenes 2 – 4.5 mm long, and two carpels per flower (Clapham, Tutin and Warburg 1962, Dandy 1980c). The coarse extreme of European material identified as *P. compressus* in collections in this country closely resembles the average of North American material on *P. zosteriformis* in vegetative character.

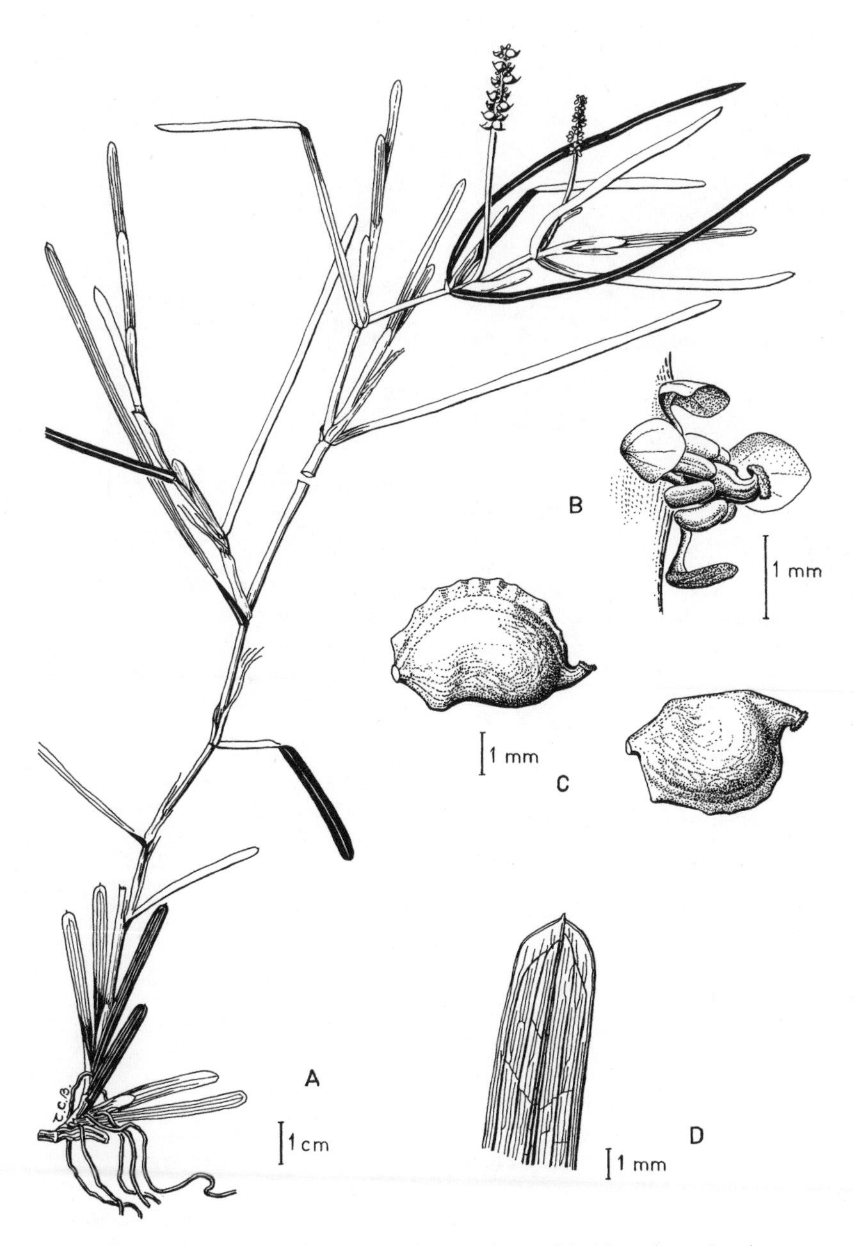

Figure 46. *Potamogeton zosteriformis*: A, base and top of fruiting plant, showing germinated winter bud at base; B, flower; C, achenes; D, leaf tip showing free-ending interneural strands.

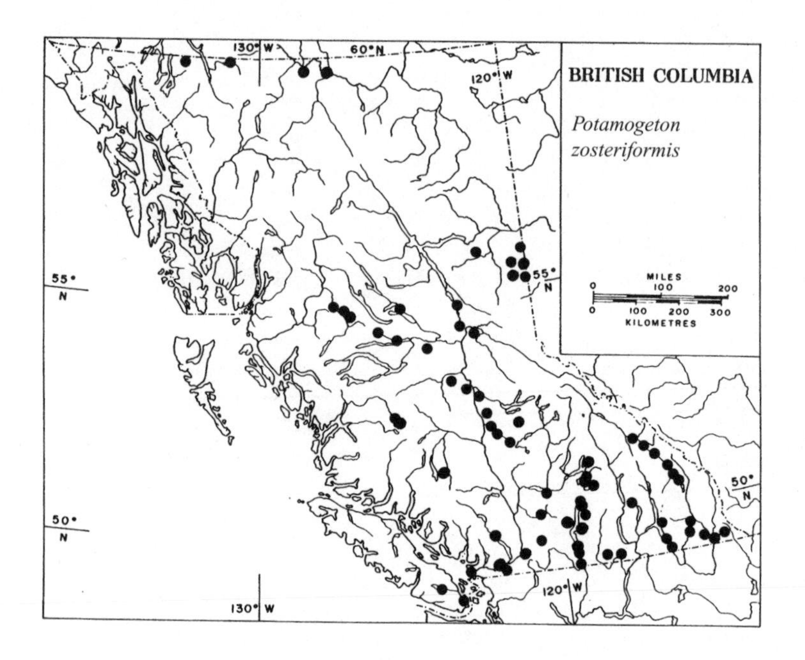
130° W
60° N
120° W
BRITISH COLUMBIA
Potamogeton zosteriformis
55° N
MILES
0 100 200
0 100 200 300
KILOMETRES
50° N
120° W
130° W

The Genus *Ruppia* L. — Ditch-grass

Rhizomatous and stoloniferous perennials with slender filiform leaves and adnate sheaths; the ligules are usually very short. Sterile plants strongly suggest *Potamogeton pectinatus*.

Leaves of vegetative shoots alternate and two-ranked, becoming subopposite at fertile branch-ends, with one vein per leaf.

Inflorescence a terminal pedunculate spike with two alternate flowers near the tip. Each flower has two minute tepals which are adnate to the divided ends of the short filaments of the two (upper and lower) stamens. Each stamen has two curved, extrorse, two-chambered anthers that together encircle the axis of the spike. Between the stamens are normally four distinct unilocular carpels on initially very short stipes, with sessile stigmas and single pendulous ovules. Pollination occurs at the water surface with floating pollen from detached and floating anthers (Tomlinson 1982).

The peduncle may elongate considerably during and after the flowering period, and subsequently may bend or coil helically, to draw the fruit beneath the water surface for ripening.

The fruits are four small achenes, barely fleshy and drupelike, from each flower. The stipes elongate appreciably as the fruit develops, and the achene body may be symmetric or asymmetric and oblique on the end of its stipe; thus each achene group presents an umbel-like appearance at maturity.

Cosmopolitan plants associated with saline or brackish coastal lagoons or marshes, and with brackish or alkaline inland lakes and marshes in drier regions.

Our material is confusingly diverse, varying in the expressions of several characters in ways that show little or no correlation. About 25 species have been described, those in this country at least, being mixed with many intergradient forms. In the absence of really clear-cut distinctions among the various forms, it is felt that the best treatment here is to maintain them as varieties of one species under the oldest specific name.

Ruppia is treated as a separate family, the Ruppiaceae, by some authors, such as Hitchcock (1969), and Douglas et al. (1994), and sometimes as a member of the Zosteraceae, as by Fernald (1950).

Ruppia maritima L. — Ditch-grass, Widgeon-grass

R. maritima* var. *rostrata* Agardh = var. *maritima

***R. spiralis* L. *ex* Dumortier; *R. maritima* ssp. *spiralis* (L. *ex* Dumortier) Ascherson and Graebner = *R. maritima* var. *spiralis* (L. *ex* Dumortier) Moris**

R. cirrhosa* (Petagna) Grande = *R. maritima* var. *spiralis

R. *occidentalis* Watson = *R. maritima* var. *occidentalis* (Watson) Graebner

This species has the characters of the genus. Varieties can be distinguished according to the following key, though some individuals do not conform to any one variety. 2n = 16, 20, 40. Figure 47.

Key to Varieties

1a. Sheaths, at least some, on main stems, 2 – 7 cm long. Leaves 8 – 30 cm long, often somewhat stiff, often in conspicuous fanlike sprays. Peduncles long and helically coiled. Achene more-or-less symmetric. Coarse stiff plants of interior lakes.............................var. *occidentalis*

1b. Sheaths all less than 2 cm long. Leaves up to 7 cm long. Fine textured, pliable plants of coastal and some interior habitats. 2

2a. Peduncles short (up to three times the length of stipe), straight or curved, but not helically coiled. Fruit asymmetric. 2n = 20. ....var. *maritima*

2b. Peduncle elongating and becoming helically coiled by the fruiting stage. 3

3a. Fruit symmetric or almost so when mature (may be asymmetric when immature). 2n = 16, 40 ....var. *spiralis*

3b. Fruit asymmetric, with a distinct offset beak. ....var. *longipes*

Variety *occidentalis* (Widgeon-grass) is probably the most distinctive variety in this province. It ranges from Alaska to Saskatchewan and Nebraska, with a generally interior distribution. Its maintenance as a species, as by Watson and others, seems justifiable though insecure, based as it is on size alone, which is susceptible to modification by environmental influences, rather than on form or shape of character-expression, which is more stable in the face of vagaries of the environment. The large, dense, dark patches formed by its colonies on the beds of shallow lakes in the interior of this province are conspicuous and recognizable even when in a vegetative state. This variety is commonly found in calcareous or slightly alkaline waters to 2 metres deep – deeper than that in which the other varieties are usually found.

The other varieties are more widespread around the northern hemisphere. Variety *maritima* extends southward to South America and Africa. Varieties *longipes* Hagstrom and *maritima*, where distinguishable as such, are confined

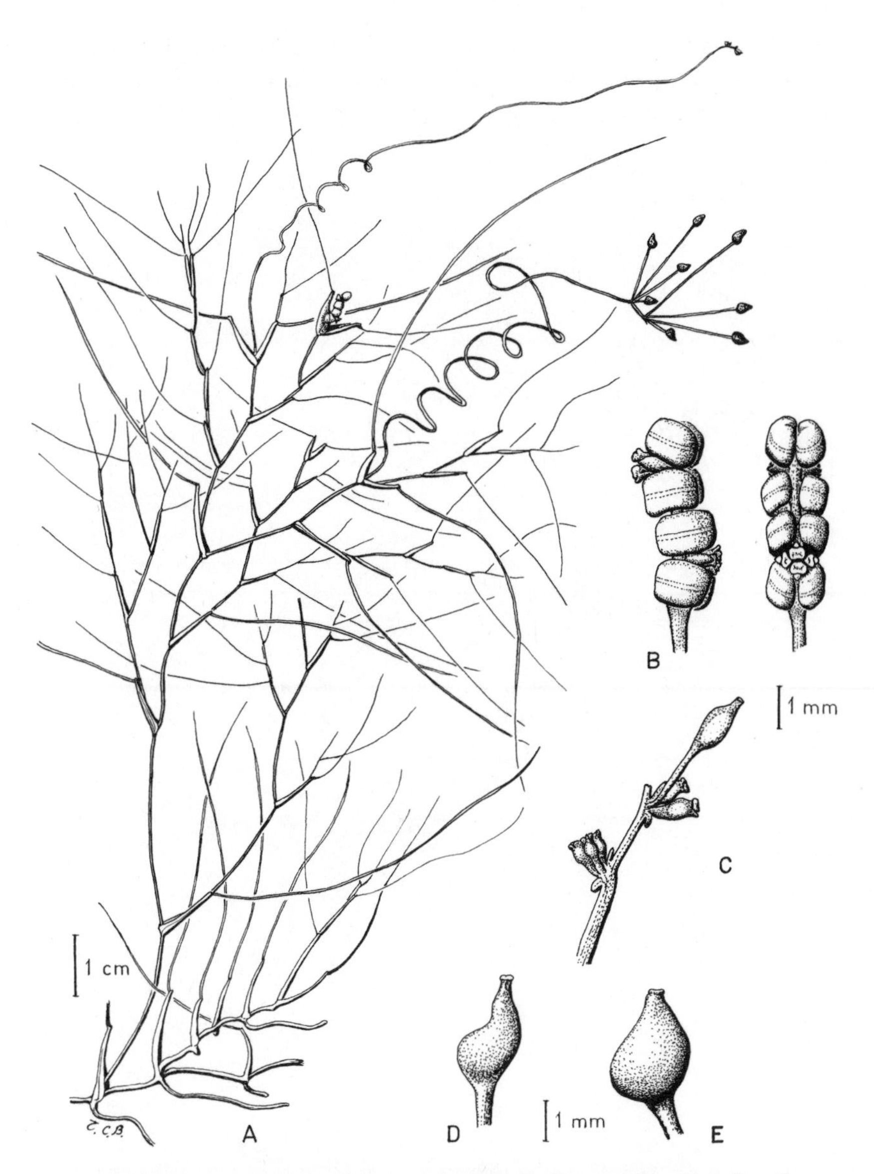

Figure 47. *Ruppia maritima* var. *longipes*: A, plant; B, flower spikes; C, spike after flowering, with one ovary fertilized; D, immature achene; E, mature achene.

to tidal estuaries and lagoons on this coast, while variety *spiralis* is found circumboreally in shallow waters of inland lakes of alkaline or brackish character, as well as in coastal lagoons.

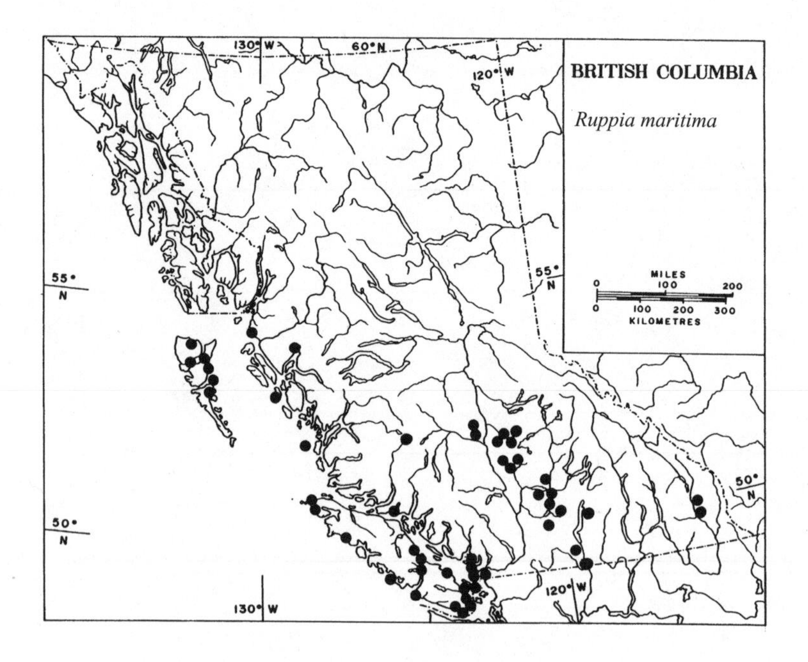

The Family Zosteraceae — Sea-grass Family

Marine perennials with creeping rhizomes rooting at the nodes and giving rise to leafy shoots with long, flexible, ribbonlike, two-ranked leaves having parallel longitudinal veins, scattered cross-veins, and intervening air channels. The tensile strength of the leaf is enhanced by the presence, beneath the epidermis, of numbers of fine longitudinal strands of sclerenchyma. Axillary scales are often present.

The inflorescence is a flat spadix sheathed in a spathe. The spadix bears sessile, unisexual flowers (according to one interpretation) in two rows on one face; and often bears a row of infra-marginal appendages beside the flowers. These appendages may be termed "bractlets", "tepals" or "retinacula" by different authors, according to their interpretation of the homologies. The exact homologies of the structures of the inflorescence and flowers in the Zosteraceae are a matter of debate (Den Hartog 1970). Similar appendages in the Old World family Aponogetonaceae are tepals, and this term is used here. The staminate flower comprises a single anther, with or without a lateral tepal. The anther comprises two sessile anther-sacs connected by a low ridge, each anther-sac containing two pollen masses. The pollen grains form long, thread-like, adhesive pollen tubes as the anthers open; and in this form they are dispersed into the water. The pistillate flower is a unilocular ovary of one carpel, with a style and two bristle-like stigmas, and contains one ovule. A lateral tepal is present in most species. Pollination is mediated by water currents beneath the sea surface. The pollen tubes, liberated in clouds, entangle with and adhere to the bristle-like stigmas of the pistillate flowers. The fruit is an achene, or a utricle that is eventually ruptured to free the single seed, which contains a curved embryo and no endosperm.

There are five genera in this family: two are primarily temperate and occur in British Columbia, and three are tropical.

Key to Genera

1a. Rhizome short, irregular in thickness. Leaves narrow but thick and opaque. Dioicous. Fruit cordate. Generally on exposed coasts. .... *Phyllospadix*

1b. Rhizome relatively elongate, more uniform in thickness. Leaves wide or narrow; if narrow, thin and often translucent. Monoicous. Fruit ellipsoid with an apical horn. Generally in sheltered bays, on soft substrata. .... *Zostera*

The Genus *Phyllospadix* Hooker — Surf-grass

Dioicous perennials with irregularly knotty rhizomes fitting the roughness of the rock to which it is attached, and bearing short stout flattish roots densely clothed with matted root-hairs. Stems flattish. Sterile shoots with short stems enclosed in leaf sheaths. Fertile shoots with longer, exposed stems.

Leaves two-ranked, ribbonlike, usually three- to five-veined, with the intervening air passages occupying half or less of the thickness of the leaf. Leaf sheaths with open, firm, pale margins that end above in rounded auricles. Leaf tips rounded, if present, but commonly becoming eroded away by surf action. Leaf blade rather thick and opaque, very dark when dry. The foliage is evergreen where not exposed for long periods at low tide.

The inflorescence is a flat spadix enclosed in the distended sheath of a leaflike bract, the blade of which tends to disintegrate before the true leaves. The spadix bears a row of infra-marginal tepals along each edge, one tepal for each flower. In the staminate inflorescence these tepals are opposite the anthers. At pollination time they spread apart and reflex, holding the margins of the spathe apart, and exposing the anthers and allowing the pollen free access to the open water. In the pistillate inflorescence the tepals alternate with the ovaries, and spread apart slightly when the fruit is formed, and the fruiting spadix protrudes from the spathe in a convex arc. The spadix has an apical, flat sterile extension that projects from the upper part of the sheath.

The staminate flower has a single anther with two separate anther-sacs. The pistillate flower has a naked, sessile ovary with a style and two linear stigmas, and contains one ovule.

The fruit is a cordate- or sagittate-based achene, beaked by the style (unless it has been eroded away), and contains one seed.

The achene, after shedding, loses its outer layer (exocarp) by decay, leaving a series of stiff, barblike cells projecting from the inner side of the central strand of each basal arm (see figure 48E). This grappling arrangement serves to anchor the achene to any fibrous substratum for germination; and may also aid in dispersal by swimming birds or mammals. On the other hand, it appears most unlikely that plant fragments, once torn loose from the rock to which they have been attached, could become reattached elsewhere in their boisterous environment. They are generally cast up on beaches to contribute to the strand lines of desiccating seaweed and driftwood.

In our species, pistillate plants outnumber staminate ones by at least ten to one. This extraordinary sex ratio may be a reflection on the efficiency of the pollination mechanism and seed production as a means of dispersal; on the other hand, it may herald the evolution of a system in which pollination is unnecessary for embryo and seed development.

Phyllospadix species typically inhabit exposed rocky shores, where they are subject to the heaviest surf. Their masses of long, ribbonlike, intensely green foliage are often seen fringing rock pools at or near the low tide level. The surf erodes and shreds the ends of the leaves so that they are seldom found intact, except when young. These leaves are able to renew themselves by growth at their bases. The bases of the plants are nearly always submerged, in rock pools or below low tide level.

The genus *Phyllospadix* consists of five species distributed along the shores of the northern Pacific Ocean: our three and two in eastern Asia.

Key to Species

1a. Stem of fertile shoot 40 – 120 cm long, with branches, and spathes often paired at the nodes. ....*P. torreyi*

1b. Stem of fertile shoot short: 5 – 40 cm long, unbranched, normally bearing a single spathe. 2

2a. Leaf with three (rarely 5) longitudinal veins, and entire margins. ....*P. scouleri*

2b. Leaf with five or, sometimes, seven longitudinal veins, and margins toothed toward the tips...*P. serrulatus*

Where they occur together, our three species tend to occupy different levels or zones. *Phyllospadix serrulatus* occurs highest in the intertidal zone, with *P. scouleri* somewhat lower, while *P. torreyi* extends deep into subtidal levels. The ranges however overlap and, in places, the species may be found growing together. Where not in competition with each other, any one of these species may invade the subtidal zone (Phillips 1979).

Phyllospadix scouleri Hooker — Scouler's Surf-grass

Plant with irregularly knotted rhizome fitting the roughness of the rock to which it is attached by short, stout, flattish roots densely matted with root hairs. Fertile stem up to 40 cm long but usually much shorter (commonly 5 – 15 cm long), unbranched, with one to four internodes and short-lived bract-leaves at the nodes, and bearing a single spathe.

Leaves up to 1.5 metres long by 1.5 – 4 mm wide, entire or rarely microscopically toothed, with three or, occasionally, five veins, narrowly rounded at apex when young. Lacunae in a single layer.

Staminate inflorescence 5 – 6 cm long. Pistillate inflorescences 4 – 5 cm long, with (9 –) 12 – 24 flowers. Mature fruit 3 – 4 mm long, cordate-based. Stem of pistillate inflorescence short at flowering time, and elongating somewhat as the fruit matures. Figure 48.

Distributed from southeastern Alaska to Baja California, Mexico, this species is generally the most common of our *Phyllospadix*. It is often a conspicuous member of the mid-tidal to subtidal plant communities on exposed rocky shores on the west coast of Vancouver Island and at a few localities on the Strait of Georgia.

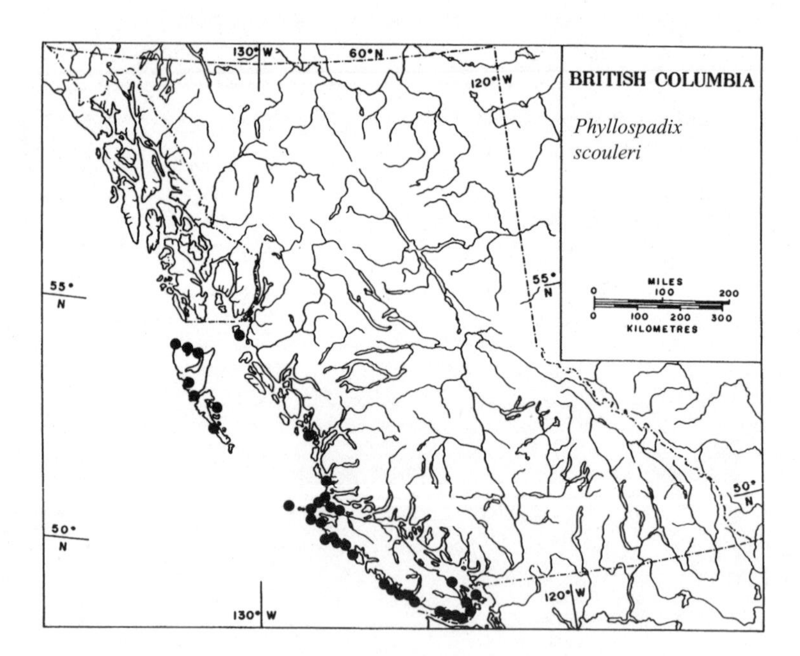

Figure 48. *Phyllospadix scouleri*: A, pistillate plant in flower: B, staminate plant in flower; C, pistillate flower; D, achenes; E, achene after decay of outer layers, showing barbed grappling hooks.

Phyllospadix serrulatus Ruprecht *ex* Ascherson

Rhizome knotty and irregular, with short internodes, as in *P. scouleri*. Fertile stems usually very short (1 – 6 cm) with one internode and no nodal leaves.

Leaves, on the average, shorter, wider (2 – 5.5 mm) and thinner than those of *P. scouleri*; with five or, sometimes, seven longitudinal veins, and margins toothed toward the apices with minute colourless teeth (soon abraded away from all but the youngest leaves). Young leaves have broadly rounded to truncate apices, which may become emarginate as abrasion wears away the extreme tip first. Lacunae are in one layer in the leaf, and very slender.

The short leafless peduncle normally bears a single spathe, which, on the pistillate plant, contains 5 – 12 flowers with very oblique tepals. Staminate plants are extremely rare – I have seen only one in many years of collecting. The few achenes I have seen are more broadly deltoid in outline than those of *P. scouleri*. Figure 49.

This species is often not distinguished from *P. scouleri*; and collections of that species often contain mixed material. Though less common that *P. scouleri*, *P. serrulatus* may be more common than existing records indicate.

Phyllospadix serulatus ranges from Oregon to southwestern Alaska; and has been found to be very abundant along the shores of Graham Island, Queen Charlotte Islands. Its ecological range is still poorly understood, at least on this coast. Local material around Victoria has been found in more sheltered rock pools and stony beaches, extending higher than, but overlapping, the range of *P. scouleri* in the intertidal zone.

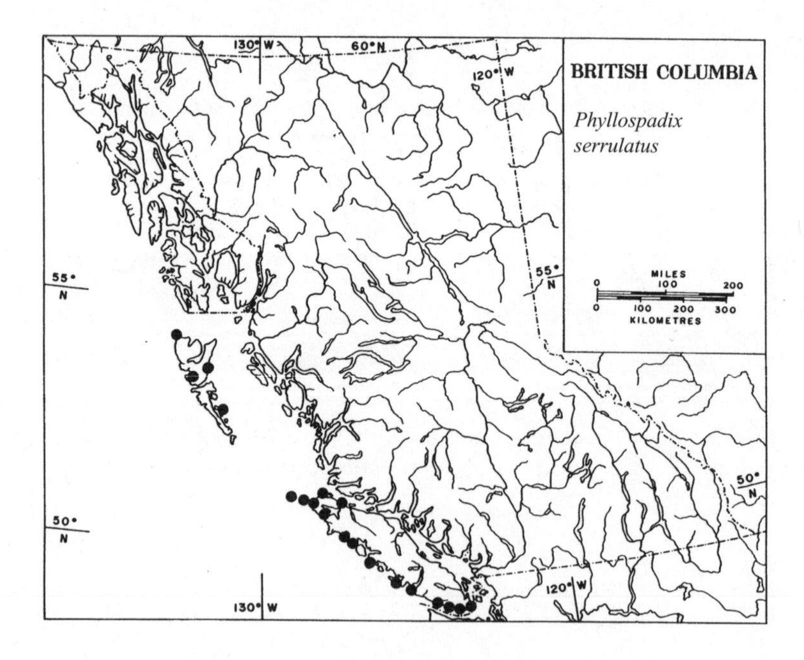

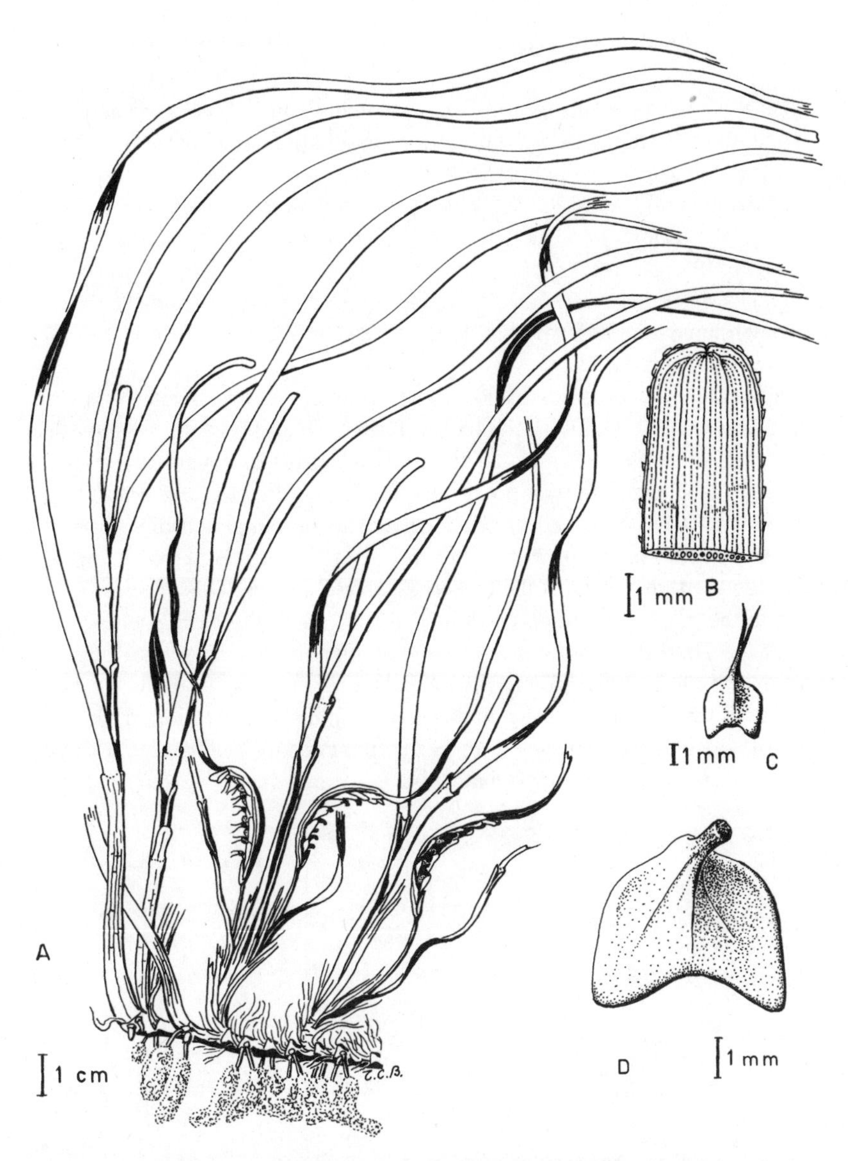

Figure 49. *Phyllospadix serrulatus*: A, pistillate plant; B, leaf tip; C, pistillate flower; D, achene.

Phyllospadix torreyi Watson — Torrey's Surf-grass

Rhizome rather more uniform in thickness than that of *P. scouleri*, but otherwise similar. Fertile stems slender, often elongating to a metre or more in length, and bearing frequent branches. Sterile shoots stemless.

Leaves generally very slender and wiry, 1.5 – 2.5 mm wide, narrower than those of *P. scouleri*, on average. Lacunae, when developed, tending to be in two layers.

Spathes several or many, one or commonly two together, terminating a short axillary branch arising from each node in the stem. Each spathe contains 14 – 20 flowers. Figure 50.

Apparently scarcer in the province than *P. scouleri*, this species is distributed from British Columbia to Baja California. I found it more common than *P. scouleri* at Triangle Island, off the northwestern tip of Vancouver Island, where it was growing in rock pools with sandy bottoms, while *P. scouleri* and *P. serrulatus* were confined to pools where they could attach directly to the rock. Our northernmost records are from Calvert Island; but it may be present along the western shores of the Queen Charlotte Islands. This species is confined to the most exposed parts of the coast, avoiding the more sheltered waters of the Strait of Georgia and adjacent straits; and characteristically grows deeper into the subtidal waters than *P. scouleri* (Phillips 1979). The relative scarcity of records of *P. torreyi* on this coast could be a reflection of the difficulty of collecting this species from its surf-pounded natural habitat. It may be more widespread and common than existing records show.

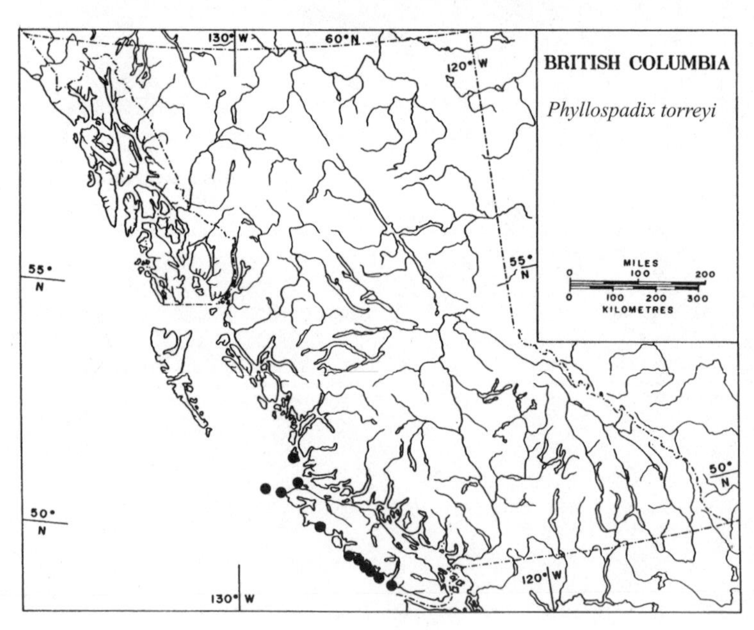

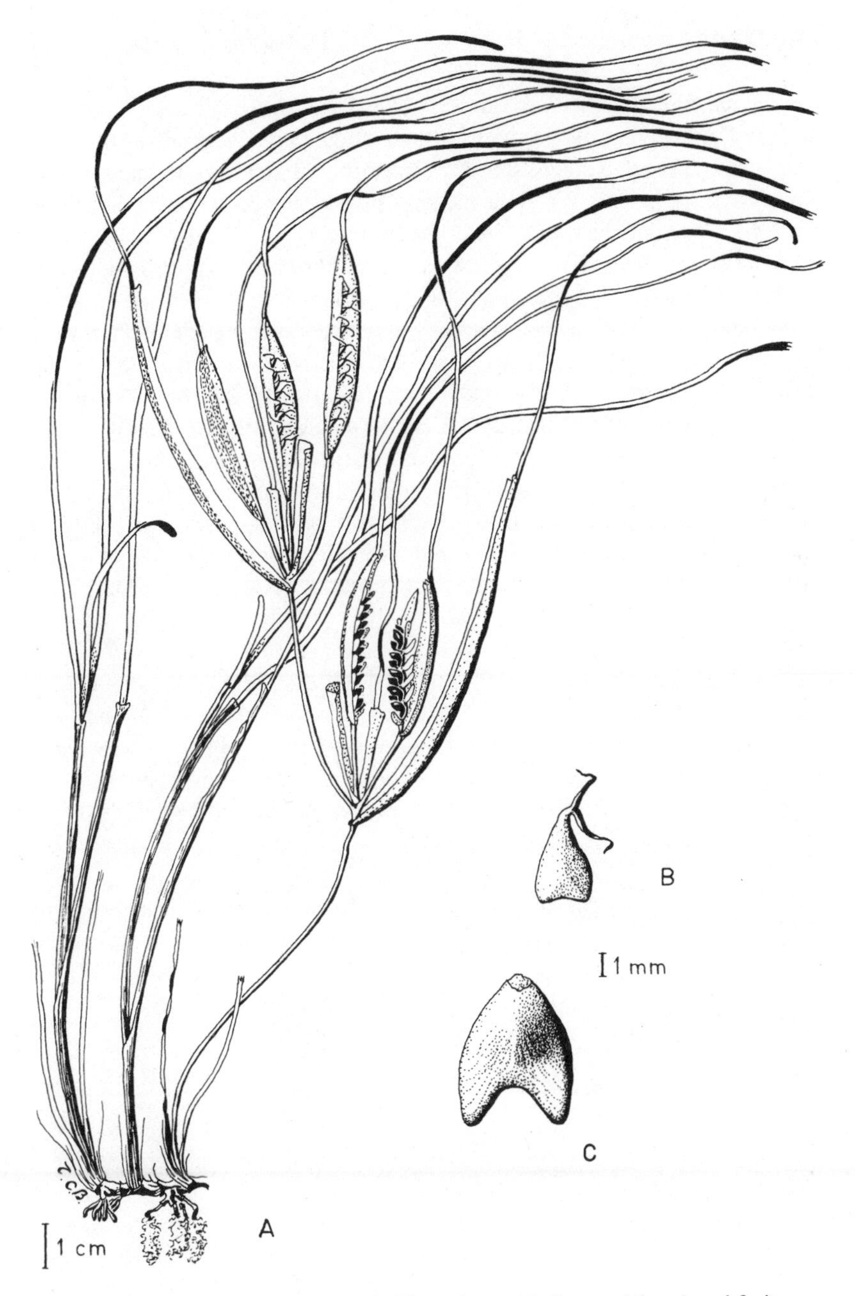

Figure 50. *Phyllospadix torreyi*: A, pistillate plant with flowers (above) and fruit; B, pistillate flower; C, achene.

The Genus *Zostera* L. — Eel-grass, Grass-wrack

Monoicous perennials with smooth rhizomes of uniform thickness, rooting at the nodes, and giving rise to shoots of two kinds. Purely vegetative shoots are short-stemmed or apparently stemless, with ribbonlike, three- to many-veined leaves rounded or emarginate at the tips, and with lacunae occupying more than half of the thickness of the leaf. Axillary scales usually two.

Fertile shoots with flat stems and supra-axillary branches, prophylls with short blades or reduced to sheaths, and leaflike bracts with blades that are shorter and narrower than the leaves of vegetative shoots. Spadix with or without minute infra-marginal tepals. Flowers alternating in two rows on one side of the flat spadix. According to one interpretation of the structure, the anthers and ovaries comprise distinct, unisexual flowers; according to another interpretation, a flower comprises one anther, of two anther-sacs, and the ovary next to it, plus, when present, the tepal adjacent to the anther. The ovary is attached at its mid-point, and has a style and two linear stigmas. The fruit is an ellipsoidal utricle, attached at mid-length, rounded at base and with an apical stylar beak. The single seed may be smooth or longitudinally ribbed.

Key to Species

1a. Sheaths on sterile shoots closed at base. Leaves 1.5 – 12 mm wide, five- to many-veined, rounded at tips. Spadix without infra-marginal tepals or with a pair at base...*Z. marina*

1b. All sheaths split to base. Leaves 1 – 1.5 mm wide, three-veined, rounded or emarginate at tips. Spadix with minute infra-marginal tepals beside the anthers. ....*Z. japonica*

Our species of *Zostera* are generally found in more sheltered waters than *Phyllospadix*, and have their rhizomes and roots buried in sediment rather than attached to rock.

Zostera japonica Ascherson and Graebner — Japanese Eel-grass

Z. americana **Den Hartog, for North American plants**

Small, very slender plant with filiform rhizomes and flowering stems, and short-stemmed or stemless vegetative shoots.

Leaves 10 – 30 cm long and 1 – 1.5 mm wide, with rounded or slightly notched tips, and with 3 longitudinal veins, the lateral two marginal; all sheaths opaque and split to the bases. The leaves soon become abraded at their tips, the first effect appearing as a notch in the initially rounded apex.

Fertile shoots lateral on the rhizome. Spadix with minute (1 – 1.5 mm long) ovate to deltoid tepals arising just within the margin adjacent to the anthers. Fruit about 2 mm long, it and the seed without longitudinal ribs. Figure 51.

Z. japonica is suspected of having been introduced with oysters from Japan (Harrison 1976). It is now found abundantly in Boundary Bay and at Tsawwassen, as well as at several points on the coast of the State of Washington. The plant occurs in silty or muddy areas, higher in the intertidal zone than *Z. marina*. I have found it in the upper intertidal zone at Tsawwassen, invading a colony of the alga *Ulva*; at Cresent Beach it now dominates appreciable areas of the tidal mud banks. It has been reported at Gabriola Island, but I have seen no material from that locality.

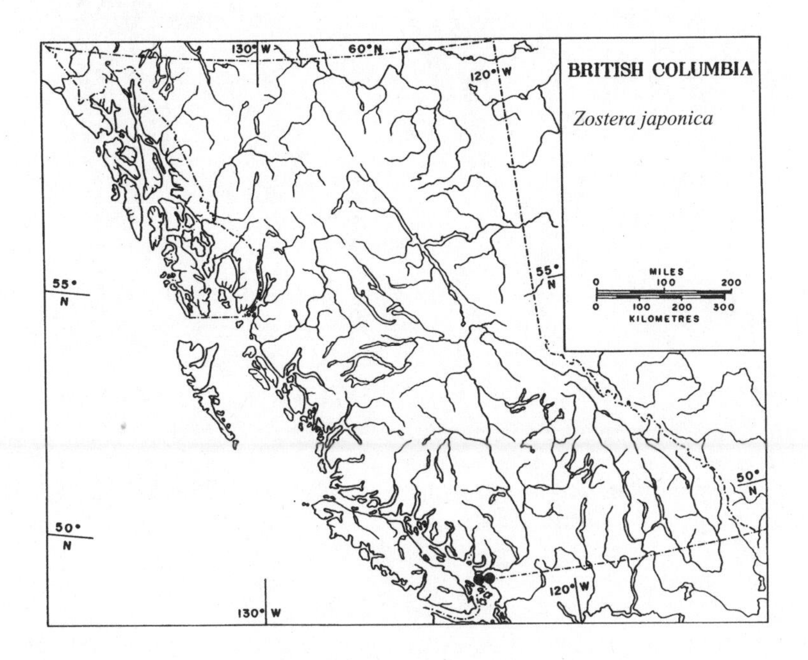

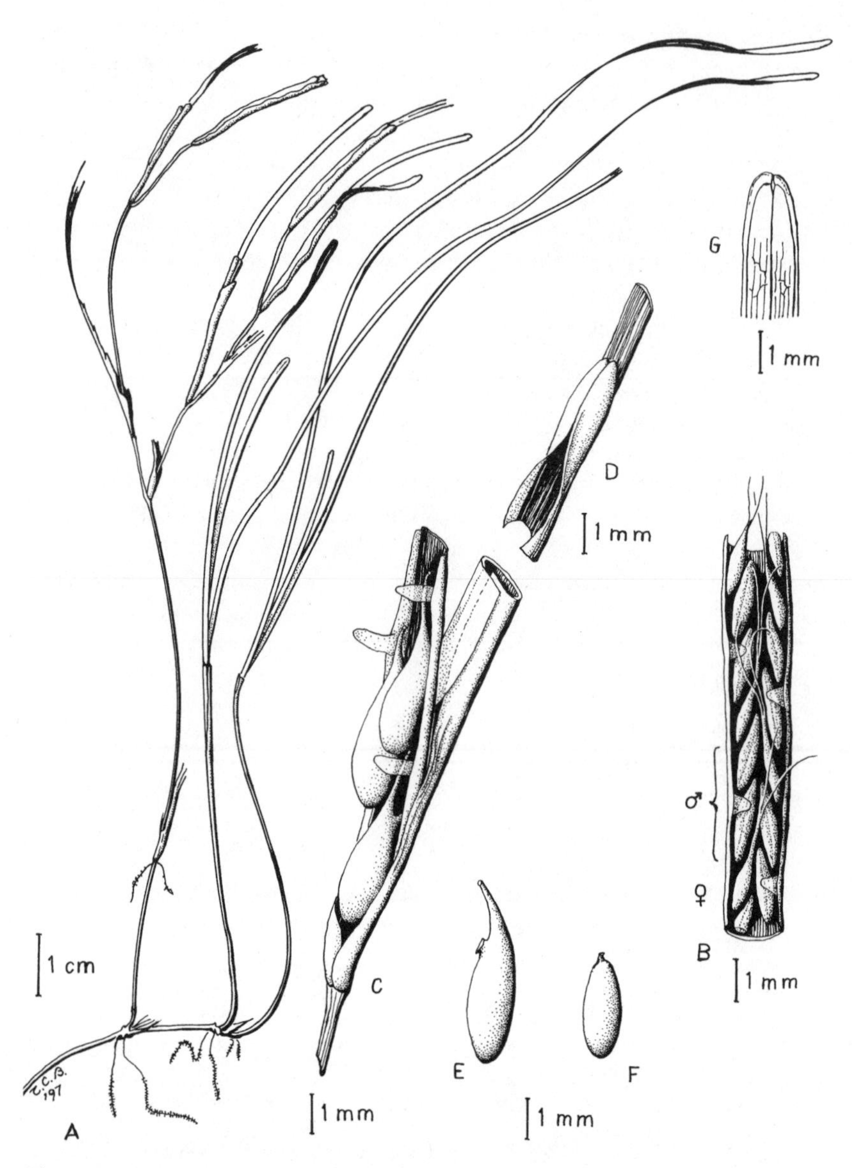

Figure 51. *Zostera japonica*: A, plant; B, segment of spadix with staminate and pistillate flowers; C, base of spathe and spadix, with fruit; D, upper part of spathe sheath; E, fruit (utricle); F, seed; G, leaf tip.

Zostera marina L. — Common Eel-grass

Z. angustifolia **(Hornemann) Reichenbach (in part),** ***Z. hornemanniana*** **Tutin, and** ***Z. stenophylla*** **Rafinesque =** ***Z. marina*** **var.** ***stenophylla*** **(Raf.) Ascherson and Graebner**
Z. pacifica **S. Watson =** ***Z. marina*** **var.** ***latifolia*** **Morong**

A plant of variable form: usually a coarse perennial, but occasionally an annual, with stout smooth rhizomes and stemless or short-stemmed, sterile, leafy branches.

Leaves variable, but generally coarse; on vegetative shoots they are often over a metre long and 1.5 – 12 mm wide, with rounded tips, and with 3 – 11 longitudinal veins, the outermost veins inframarginal. The sheaths of vegetative shoot leaves are closed, flat-tubular, closely enclosing the shoot, and ventrally translucent.

Fertile shoots are terminal on the rhizomes, long-stemmed, up to two metres or more long, with supra-axillary branches and with leaves shorter and narrower than those of the sterile shoots. Prophylls with or without short blades. Sheaths of the spathes open to base, with firm, opaque margins. Spadix without tepals or with one each side beside the lowest anther.

At flowering time the two elongate tapering stigmas of the pistillate flowers project from between the overlapping margins on the spathe. The pistillate flowers are receptive before the anthers in the same spathe are ripe; and the stigmas are shed by the time the pollen of adjacent anthers is discharged. Pollen grains are threadlike, tubular, and up to 3 mm long at the time they are discharged. They are discharged into the water in clouds from ripe anthers and are carried away on water currents to entangle with and adhere to the bristle-like stigmas of receptive pistillate flowers.

The fruit is a utricle, 5 – 8 mm long, centrally attached, with a hornlike stylar beak at one end. The utricle ultimately ruptures to expose the seed, which is 2.5 – 4 mm long, and longitudinally ribbed. Fruiting shoots ultimately become detached and drift away, commonly becoming washed up on beaches with seaweeds. 2n = 12. Figure 52.

Key to Varieties

1a. Leaves of sterile shoots 1 – 6 mm wide, opaque, with three or, sometimes, five veins. Stems filiform. .... var. *stenophylla*

1b. Leaves of sterile shoots 4 – 12 mm wide with five to eleven veins. Rather translucent. Stems robust. 2

2a. Leaves of sterile shoots 4 – 10 mm wide with five to seven veins. .... var. *marina*

2b. Leaves of sterile shoots 10 – 12 mm wide with seven to eleven veins, translucent. .... var. *latifolia*

Widely distributed along sea coasts of the northern hemisphere, *Zostera marina* often forms extensive meadowlike colonies in sheltered lagoons and bays. It roots in a bottom of coarse or fine sediment, seldom being found on rock. The base of the plant is usually permanently submerged, sometimes by a few metres. In British Columbia, *Zostera marina* occurs generally along the coast; the absence of records from the central coast being an artifact, and representing an absence of collectors rather than an absence of the species.

It has been suspected, though not demonstrated, that the above varieties may represent merely individual responses to variation in such environmental factors as water temperature or degree of shelter or exposure (Den Hartog 1970).

Variety *stenophylla* has been recorded from the Queen Charlotte Islands, Nanoose Bay on Vancouver Island, Boundary Bay and Savary Island. The Queen Charlotte Island material was intertidal in situation; but whether this is a general property of this variety cannot be confirmed. Variety *latifolia* is subtidal – only the upper parts of the plants reach the surface. It has been recorded along the Pacific coast of North America from southern British Columbia to California. In British Columbia, it has been found in sheltered waters, as in Sooke Basin and Bamfield Harbour. An annual form, producing fertile shoots directly from seeds, without overwintering rhizomes, has been found where this species invades the upper intertidal zones on the coast of Oregon (Bayer 1979); it has also been reported on the Altlantic coast (Bayer 1979). At the upper levels, lengthy exposure, temperature extremes and disturbance by grazing waterfowl are postulated as possible causes for the failure of rhizomes to become established and to survive over winter.

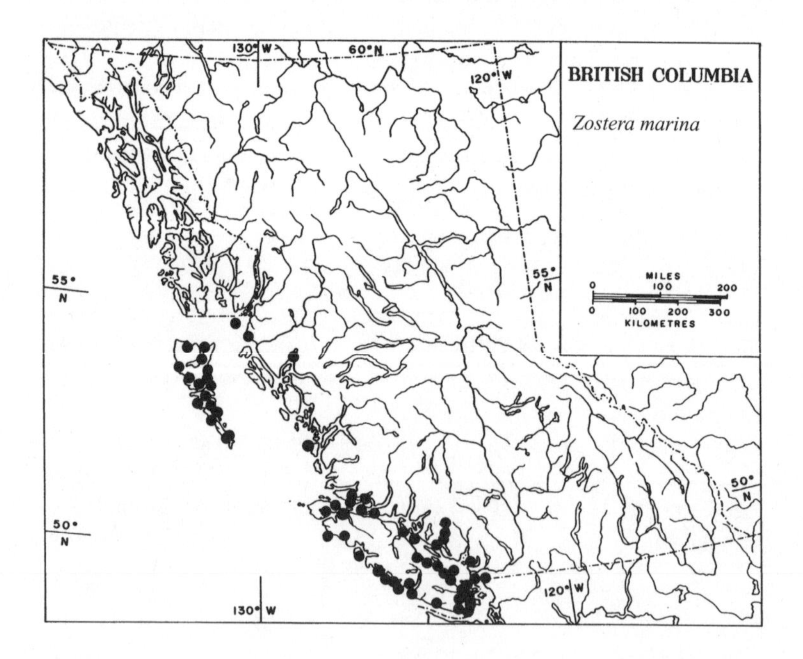

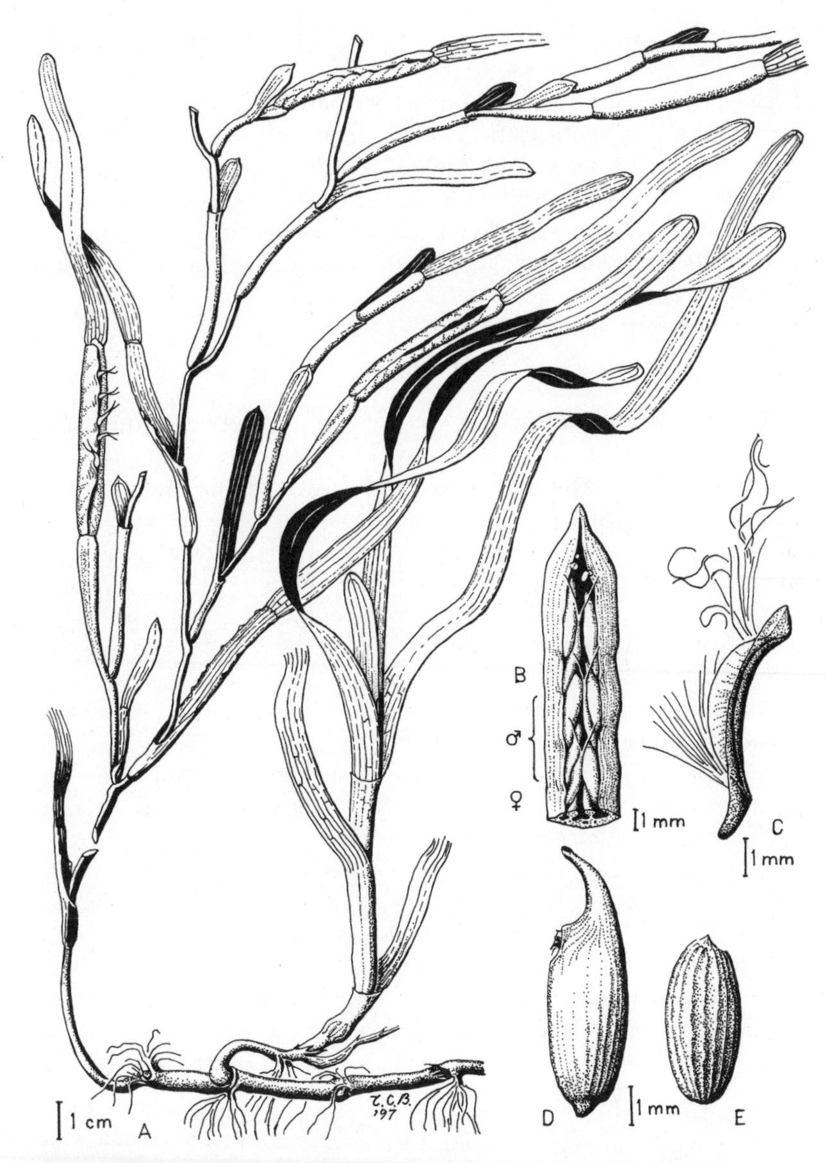

Figure 52. *Zostera marina*: A, plant, with vegetative and flowering branches; B, upper part of spadix, with staminate and pistillate flowers; C, anther shedding pollen; D, fruit (utricle); E, seed.

The Family Zannichelliaceae — Horned Pondweed Family

Submerged rhizomatous or stoloniferous perennials with linear, alternate or subopposite leaves and sheaths. Flowers unisexual, sometimes dioicous. Staminate flower has one to three stamens, sometimes with a rudimentary perianth (none in ours). Pistillate flower has two to nine separate carpels, each with a single pendulous ovule. Fruit an achene, stipitate in ours, containing a seed without endosperm. The family contains three genera; only one of them in North America.

The Genus *Zannichellia* — Horned Pondweed

Submerged perennial with often twining roots, slender rhizomes or stolons, and filiform leafy stems. Leaves alternate or subopposite, apparently in four ranks, narrowly linear to filiform, with one central vein, and with sheaths free from the leaf bases.

The pattern of stem growth involves a cycle of three successive internodes, only the first two of which elongate. Thus, of the three two-ranked alternate leaves in the cycle, the first, which may be bladeless and represented by only a sheath, is obviously alternate, while the remaining two form a subopposite pair. Growth from axillary buds repeats the cycle, but with the ranks at right angles to those below, producing a four-ranked appearance. Above the third leaf, the stem terminates in a pistillate flower (Posluszny and Sattler 1976a).

Flowers monoicous, unisexual, greatly reduced and lacking perianths in our species, terminal though appearing axillary due to prolonged vegetative growth from adjacent axillary buds. The staminate flower is a single stamen terminating a short branch in the axil of the lower leaf of a subopposite pair. The pistillate flower consists of two to five – commonly four – separate carpels with oblique, funnel-like stigmas. Pollination takes place under water. The fruit is a stipitate achene tipped with a stylar beak.

There are two species of *Zannichellia*: one in Africa and the one described below.

Zannichellia palustris L. **Horned Pondweed**

Freely branching, slender, rhizomatous or stoloniferous perennial up to 50 cm or more long. Leaves 2 – 10 cm long by 0.5 – 1 mm wide, with flaring sheaths free from the leaf bases and encircling the stem. The bladeless sheath that represents the first leaf of the stem cycle initially encloses the entire nodal complex of the shoot apex. In the axil of the lower leaf of the subopposite pair, an axillary branch terminates in a staminate flower consisting of a single stamen with two or four anther-sacs, and a protruding connective. The terminal pistillate flower, beside the axillary bud that gives rise to prolongation of growth, is short-stalked, surrounded by a perianth-like sheath, with four (or rarely two) carpels on short stipes, with prominent styles 1 – 1.5 mm long terminating in oblique, peltate or funnel-like stigmas. Achenes spreading on widely diverging stipes, oblong, two or three times as long as their thickness, 2 – 4 mm long, with crested and toothed dorsal keels, and with prominent stylar beaks 1 – 2 mm long, with more or less hooked tips formed from the stigmas. 2n = 12, 24, 28, 32, 36. Figure 53.

A nearly cosmopolitan species, *Zannichellia palustris* is widespread in North America. It reaches northward to coastal Alaska, to 66°N latitude on the Mackenzie River in the Northwest Territories (Cody 1998), to James Bay and Newfoundland. In British Columbia it has been found mainly in the southern and central regions, including Vancouver Island, and northward as far as Fort St John in the Peace River basin; and is probably more widespread.

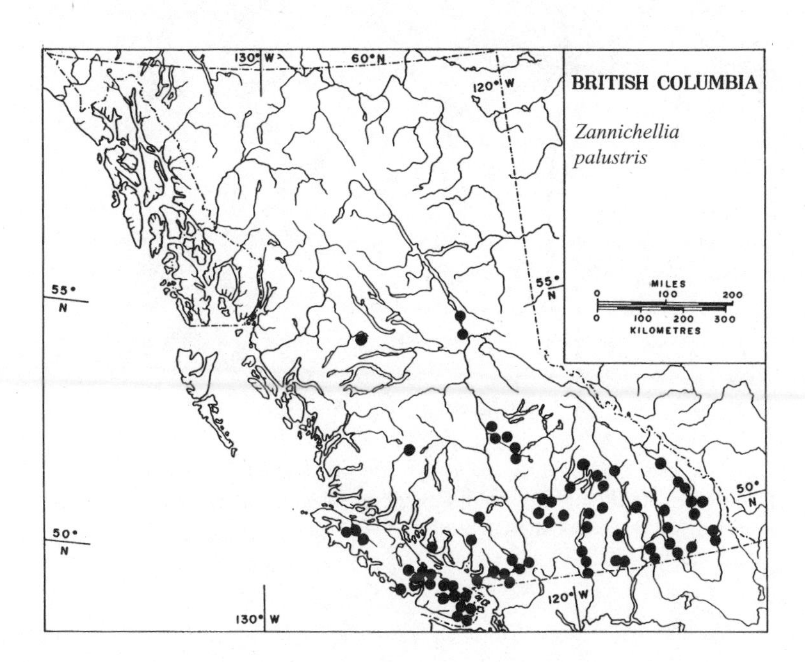

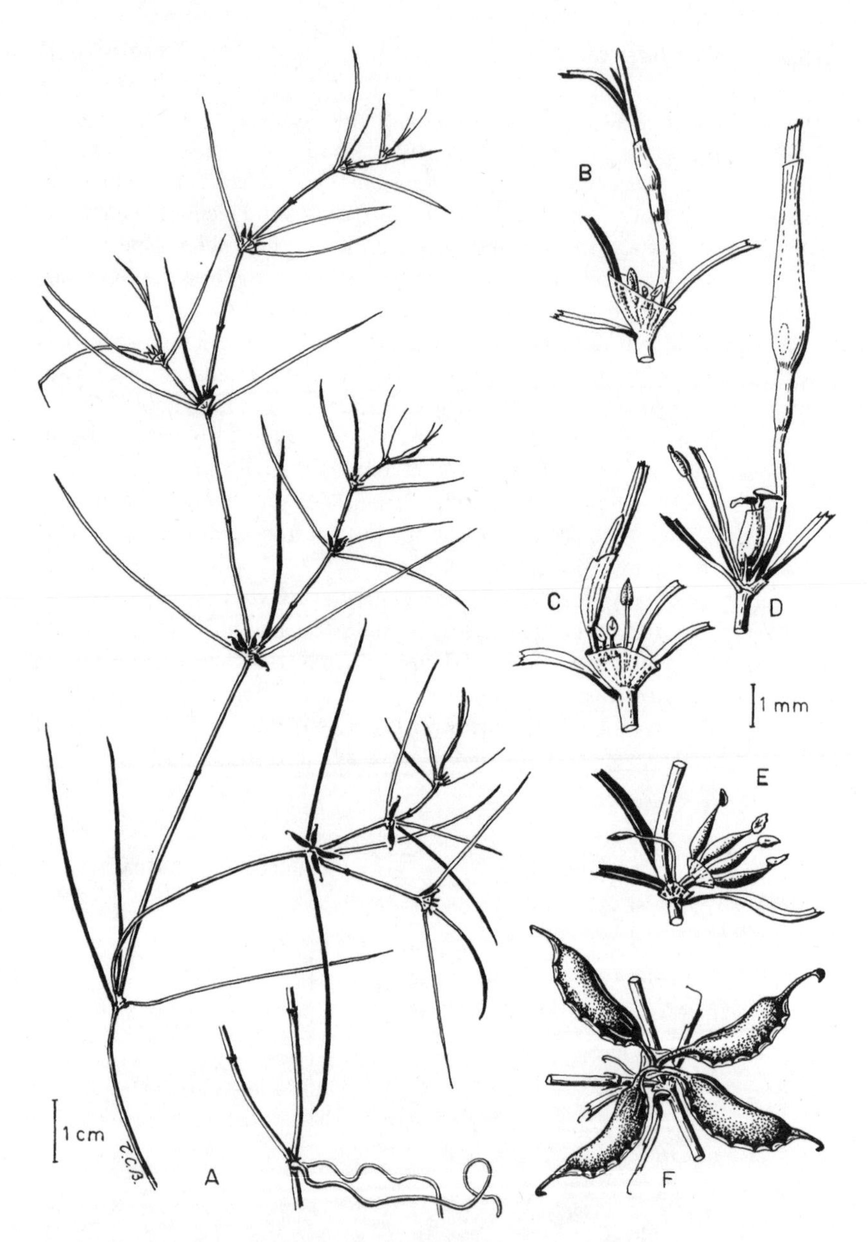

Figure 53. *Zannichellia palustris*: A, plant; B, young shoot tip, with immature staminate and pistillate flowers in sheath; C, node and sheath with later stage of flower development; D, node with mature flowers; E, developing fruits; F, mature fruits (achenes).

The Family Najadaceae — Naiad or Water Nymph Family

Mostly annuals, with subopposite, four-ranked (or occasionally falsely whorled) linear leaves with single veins and dilated sheathing bases with axillary scales.

Flowers solitary or paired in the leaf axils, of very unorthodox development (Posluszny and Sattler 1976b), unisexual. The staminate flower is a single subsessile anther enclosed in two membranous envelopes with lobed orifices, the inner envelope becoming adnate to the anther. The pistillate flower starts as a single ovule terminal on the floral axis, and becomes enclosed by a pistil with a slender style and two to four linear stigmas. The outer envelope around the stamen and the pistil around the ovule are identical in form and development. Pollination occurs beneath the water surface.

The fruit is a membranous-walled achene tipped by the persistent style and stigmas, and filled by an elongate seed with a straight embryo and no endosperm.

The Genus *Najas* L. — Naiad

The only genus, comprising about 40 species; one in British Columbia.

Najas flexilis (Willdenow) Rostkovius and Schmidt — Water Nymph or Naiad

Submerged monoicous plant, 10 – 150 cm long, profusely branched with alternate branches, and having a tufted appearance. Leaves in subopposite pairs, 1 – 3 cm long by 0.5 – 1 mm wide, gradually tapering to slender tips, finely toothed along the margins and with dilated, toothed, sheathing bases with paired axillary scales.

Flowers sessile or subsessile in the axil of the lower of the pair of subopposite leaves; single or a staminate and a pistillate flower together. Anther with one anther-sac. Achene 3 – 4 mm long with a stylar beak 1 – 2 mm long, tipped by three stigmas. Seed smooth and shiny, 3 mm long by 1 – 1.5 mm thick. 2n = 12, 24. Figure 54.

Najas flexilis is widespread across North America and in northwestern Europe. In British Columbia it has not been recorded north of 55° North latitude, though it reaches the northern border of Alberta. It grows submerged in clear, shallow (0.5 metre) to moderately deep, fresh to brackish water. Though typically an annual, this plant has been reported occasionally to survive over winter as the basal part of the plant, which produces new buds and shoots in the spring (Stuckey, Wehrmeister and Bartoletta 1978).

Where abundant, this is an important plant for waterfowl, which eat all parts of the plant (Fassett 1957).

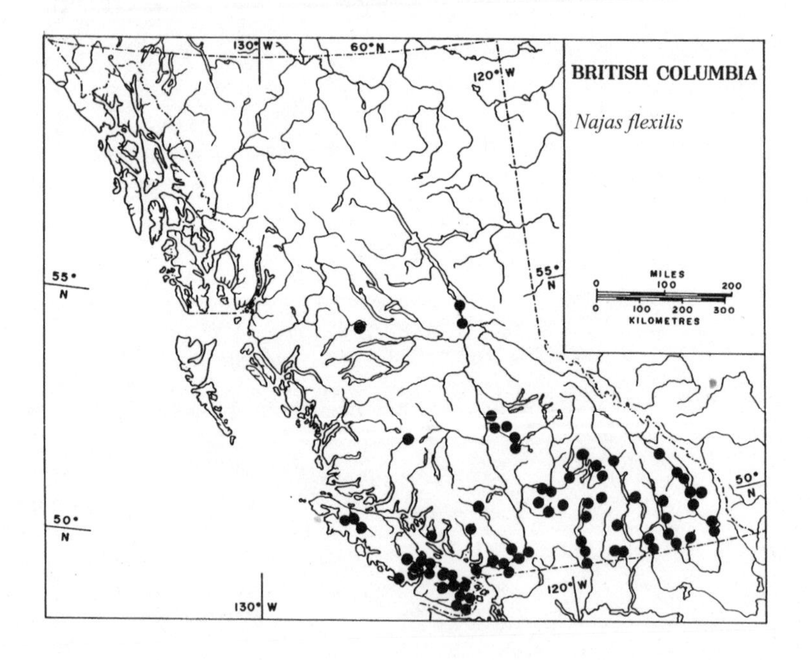

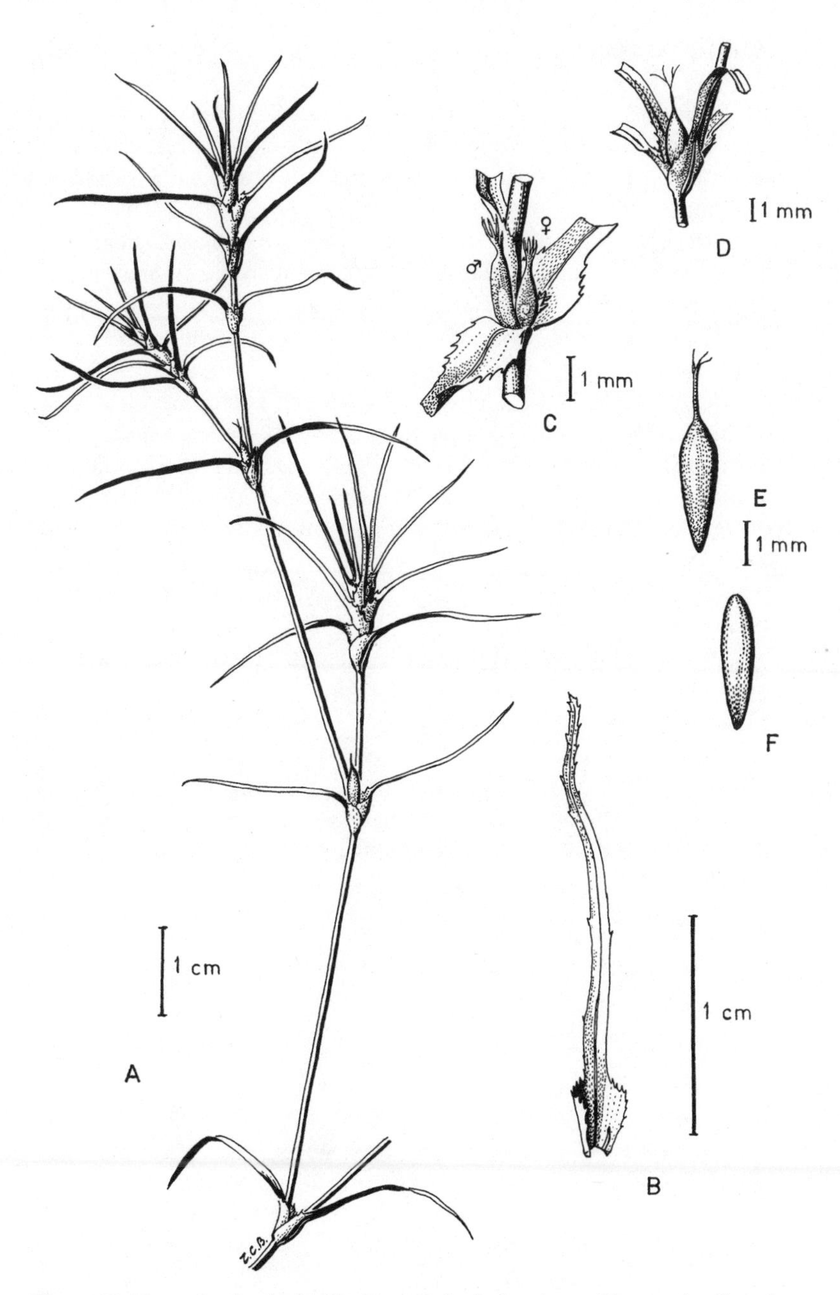

Figure 54. *Najas flexilis*: A, fertile shoot; B, leaf, showing axillary scale; C, node, with staminate and pistillate flowers among the leaf bases; D, node with fruit; E, fruit (achene); F, seed.

The Family Araceae — Arum Family

Perennial herbs (or vines, shrubs, or even small trees in the tropics), commonly with acrid juice and bundles (raphides) of calcium oxalate crystals in many of the cells, stout stems or rhizomes, and commonly with two-ranked leaves with sheathing bases.

The inflorescence is a densely flowered, axillary, often fleshy spadix, subtended and, at first, usually enclosed by a conspicuous coloured spathe. Individual flowers small, bisexual, or unisexual and monoicous, the ovaries and floral bases sometimes embedded in the fleshy body of the spadix; sometimes some flowers reduced to sterile appendages.

Fruit generally fleshy, either a cluster of berries or a compound fruit composed of the fleshy spadix and the ovaries and seeds that are embedded in it.

A large family, of about 110 genera and 1800 species, mostly of humid tropical and subtropical regions. While the species in other parts of the world are generally terrestrial or even epiphytic, the few species in British Columbia are aquatic or amphibious.

Key to Genera

1a. Leaves and spathes linear. Leaves set edgewise to the stem. Spathe green, never enclosing the spadix. ....*Acorus*

1b. Leaves and spathes broad. Spathe at first wrapped around the spadix. 2

2a. Leaves net-veined, lanceolate to elliptic, very large, forming a rosette. Spathe yellow in ours. ....*Lysichiton*

2b. Leaves pinnately closely parallel-veined, heart-shaped, two-ranked. Spathe white...*Calla*

The Genus *Acorus* L. Sweetflag, Calamus-root

Aromatic plants with horizontal rhizomes and erect flowering stems; the two-ranked leaves without distinction between petiole and blade, set edgewise to the stem but sheathing at the bases. Inflorescence with a long peduncle arising from the rhizome; the spathe similar to a leaf, green prolonged beyond the obliquely set spadix, which it never encloses. The tissues without raphides (an exception to the family characters in both these last characteristics). Flower minute, with six tepals and six stamens, greenish to yellowish; the ovary three-loculed, with several ovules. Fruit a spike of tiny dry capsules (an exception to the family character in this respect), with three or more seeds in each capsule.

A circumboreal genus comprising two or three species, depending on interpretation.

Acorus americanus (Rafinesque) Rafinesque — American Sweetflag

A. calamus **L. var.** ***americanus*** **Rafinesque**
A. calamus **of ed. 1, not of Linnaeus**

Leaves up to a metre long by 2 cm wide, shiny yellowish green, with their edges toward the peduncle above the sheathing bases, producing an irislike aspect.

Spadix projecting obliquely upward from the top of the peduncle, greenish to yellowish; the erect, green leaflike spathe projecting up the 80 cm beyond the attachment of the spadix, to which it is set edgewise; and looking like a leaflike extension of the peduncle. Mature spadix up to 7 cm (rarely more) long by 1.5 – 2 cm thick.

Fruit a dry capsule with three locules. The few fruiting specimens I have examined contained one seed per locule. 2n = 24. Figure 55.

Acorus americanus ranges from the central interior of British Columbia eastward across Canada to Newfoundland, and from Alaska and the southwestern Northwest Territories southward to the states of Washington and Texas, including some widely scattered localities in the interior of British Columbia. It is found in marshy areas, commonly in shallow water (up to 0.75 metre deep), and may invade cat-tail (*Typha*) colonies.

Packer and Ringius 1984 have shown that the common species of this genus occurring over most of North America is a native species, and is distinct from *A. calamus*. *A. calamus* is a sterile triploid (3n = 36) clone cultivated in the Old World, propagated by division of the rhizome and introduced into eastern North America. Only *A. americanus* is thus reported from British Columbia.

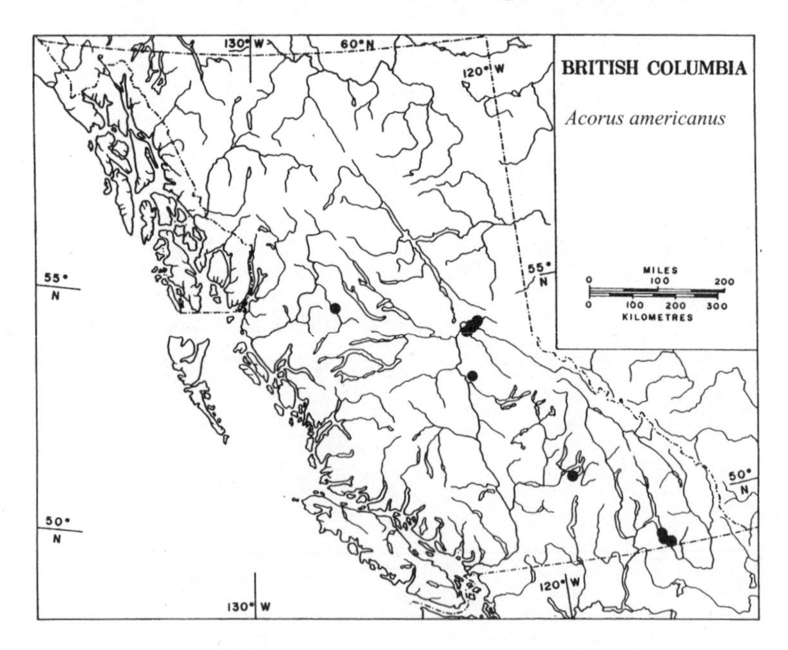

These two entities are almost indistinguishable by gross morphology; though the mature spadix of *A. calamus* is generally somewhat longer and thinner (commonly 7 – 9 cm long by about 1 cm thick) than that of *A. americanus*. Other distinctions are based on the chromosome number, the occurrence or absence of mature seeds, and the viability, or lack of it, of the pollen grains, as revealed by chemical stains.

All the British Columbia material in the collection of the Royal B.C. Museum has been identified by Sue A. Thompson as *A. americanus*.

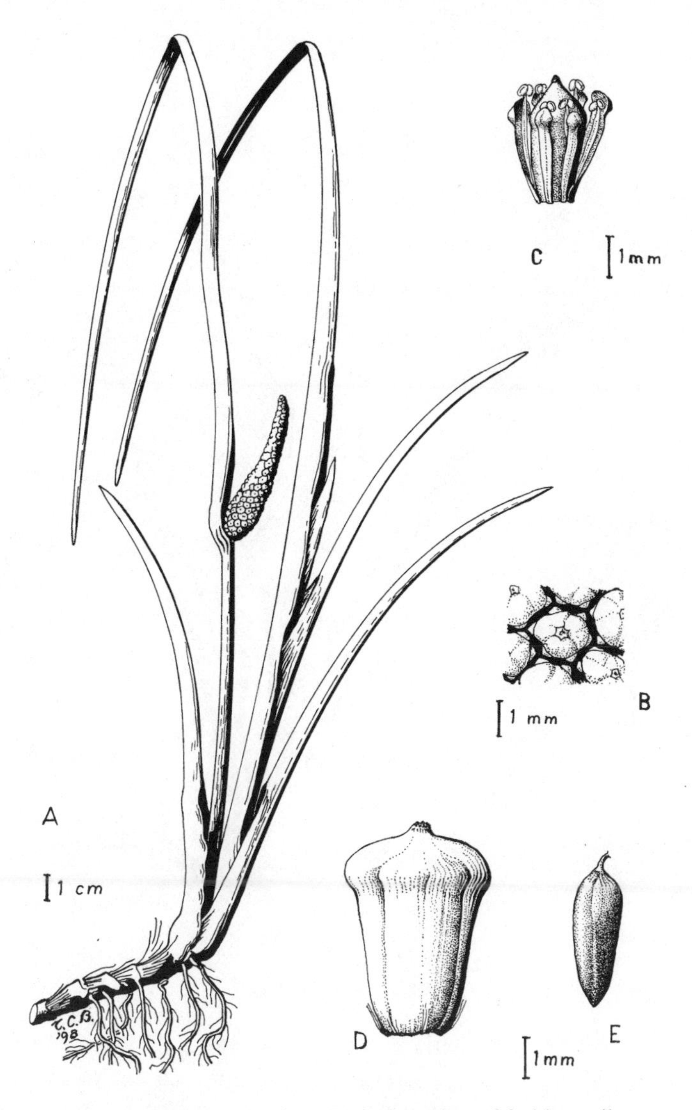

Figure 55. *Acorus americanus*: A, plant; B, surface view of fruiting spike; C, flower; D, fruit (capsule): E, seed.

The Genus *Calla* L. — Calla

Low, trailing perennial plants with two-ranked leaves. Flowers without perianths; the lower ones on the spadix bisexual, with six stamens and a superior, unilocular ovary; the upper flowers consisting of stamens only. Fruit a cluster of berries, each with few to several seeds. Spathe opening wide.

Only the following species.

Figure 56. *Calla palustris*: A, flowering plant: B, fruiting spadix and spathe.

Calla palustris L.

Plant with long, horizontal, often floating or submerged stems, the nodes giving rise to many long roots, and to alternate, two-ranked leaves.

Leaves with heart-shaped, pinnately parallel-veined blades that have acuminate tips, and with petioles that have ample ligular sheaths.

Spathe wide open, spreading, white, normally single, but rarely double or triple in forma *polyspathacea* Victorin and Rousseau. Flower spike white by reason of the crowded white anthers; the lower part of the spike punctuated by the projecting green pistils of the bisexual flowers. Berries red when ripe. 2n = 36, 72. Figure 56.

This species of circumboreal distribution ranges across North America from Alaska to Labrador, reaching the Arctic coast near the mouth of the Mackenzie River. It is found mainly in the Boreal and Subalpine forest regions (Rowe 1972) in British Columbia. The plants are commonly seen creeping or floating in cold northern swamps, fens and marshy lake margins, often among emergent shrubby vegetation, with at least some of its roots attached to the shore or to some solid substratum.

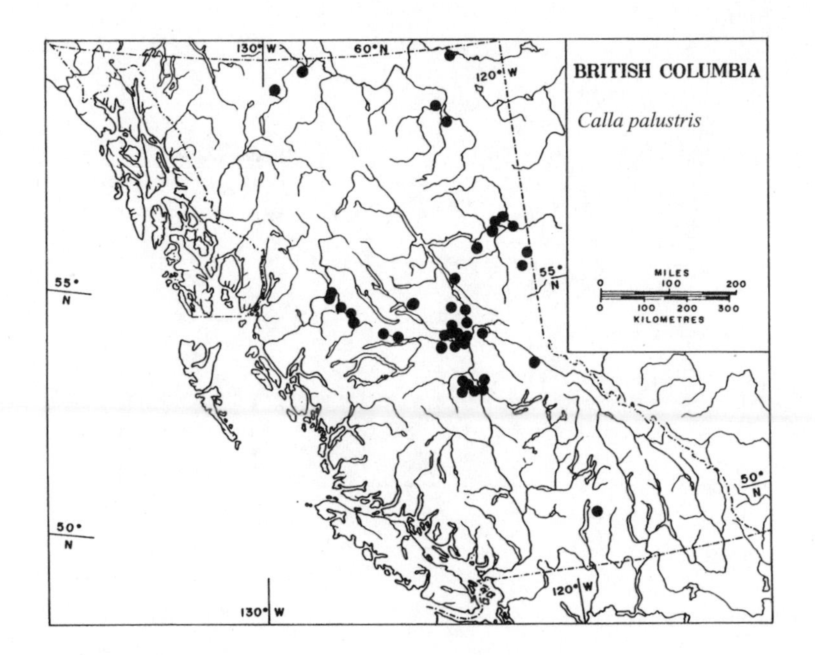

The Genus *Lysichiton* Schott

Coarse perennials with large net-veined leaves. Spathe large when open, forming a hood around the spadix. Flowers all alike and complete, with four tepals, four stamens, and a two-loculed inferior ovary largely sunk in the fleshy spadix body. Compound fruit formed of the rather fleshy spadix, with the many ovary bases, their locules and seeds, embedded in the compound mass.

This genus has two species: ours and one in eastern Asia. The name is sometimes spelled "*Lysichitum*".

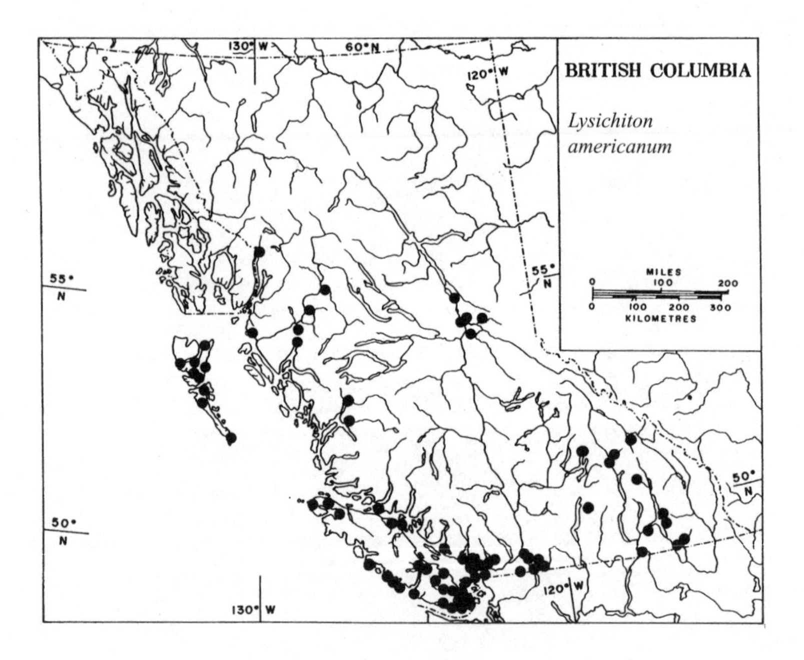

Lysichiton americanum Hulten and St John — Skunk Cabbage, Yellow Arum

Rhizome short, thick, ascending, producing leaves in large rosettes. Leaves up to 1.5 metres long and 0.5 metre wide, thin, net-veined, acute to rounded at the apex, tapering downward to usually short, winged petioles. The plant, especially when flowering, has a skunklike smell, but becomes pleasantly aromatic on drying.

The inflorescence appears before the leaves in spring, and is later overtaken by their growth. Spathe yellow, conspicuous, decaying after flowering is finished. Flower mass yellow at pollinating time, when the smell becomes most noticeable. Pollination is effected by small beetles. After the flowering season, the spathe withers away and the fruiting spadix is naked, pale green and rough-surfaced. Figure 57.

Plants of the western slope of Vancouver Island often have leaves with dark bluish green blotches, and pale yellow spathes; they may flower when the leaves are comparatively well grown.

Ranging from southern Alaska to California, and eastward to Montana, this species is widespread in British Columbia from about latitude 55° N southward. Verbal reports of its occurrence in the Atlin region, though not substantiated by actual specimens, are yet believable; since the species occurs along the Alaskan coast northward to Prince William Sound, and westward to Kodiak Island. It could conceivably cross the border along the Taku River to reach the area of British Columbia south of Atlin. It is commonly found in swamps, wet forest and wet meadows, especially where shaded; and may be seen standing in water in late winter or spring.

Figure 57. *Lysichiton americanum*: A, plant at late flowering stage: B, fruiting spadix.

The Family Lemnaceae — Duckweed Family

Tiny floating or submerged plants lacking any clear distinction between stem and leaf. Rootlets present or absent, bearing root-caps but no root hairs. Raphides are found in some species, and axillary scales in a few species. Xylem tissue is commonly replaced by lacunae, and where it occurs, it is unlignified.

Flowers minute, two or three together in a small pouch, normally one pistillate flower accompanied by one or two staminate flowers. No perianth. Staminate flower a single stamen. Pistillate flower a single carpel with a short style, and containing one to several ovules. Fruit is a one-seeded utricle.

The world's smallest flowering plants, members of this family display the ultimate reduction of a flowering plant in specializing for a floating aquatic life. Reproduction is almost entirely vegetative. Flowers of most species are seldom found, and for some they have never been found. Vegetative reproduction is by budding of young plants directly from pockets in the edges or basal ends of older plants. By this means these plants can colonize small ponds and ditches very rapidly, producing, in one season's growth, extensive floating masses of many thousands of tiny plants. At least some species produce minute, rootless bulblets. These sink to the bottom for the winter, and serve to regenerate the species vegetatively the following spring.

The Lemnaceae are a small worldwide family of four genera and 28 species. They are generally believed to be profoundly reduced relatives of certain small floating members of the Araceae (Daubs 1965). Some species typically occur on eutrophic or polluted waters of limited extent.

Members of this family are eaten by waterfowl, and plants can be dispersed by the feet or feathers of these birds.

Key to Genera

1a. Roots absent. Plant body globose to ellipsoid, about 1 mm across. .... *Wolffia*

1b. Roots present. Plant body flat, larger.......... 2

2a. One root per plant body. .... *Lemna*

2b. Several roots in a tuft. .... *Spirodela*

The Genus *Lemna* L. — Duckweeds

Plants of various shapes, usually flat; each plant with one root hanging beneath it, and with two marginal reproductive pockets.

Flowers three together – one pistillate and two staminate – initially in a sheathing spathe, all in a marginal pocket. Anther two-chambered. Ovary with one to seven ovules. Fruit usually one-seeded.

Lemna is an almost worldwide genus of about ten species.

Key to Species

1a. Plant body broadly elliptic or nearly circular. Sessile or nearly so where attached to its parent plant. ....*L. minor*

1b. Plant body narrowly obovate to lanceolate, distinctly stalked. ....*L. trisulca*

Lemna minor L. *sensu lato* — Common Duckweed

Including *L. turionifera* Landolt

Plant disclike, broadly elliptic, 2 – 5 mm long, green and smooth above, often purplish beneath, obscurely three-veined, or seldom five-veined, with a single root attached centrally to the underside. Young plants stalkless or very short-stalked, produced from a pair of marginal pockets near the basal end of the parent plant.

Fruit broadly flask-shaped, tipped with a short style, less than a millimetre long. The plant survives over winter as small rootless tubers on the bottom of the pond. 2n = 20, 30, 40, 50 (Gleason and Cronquist 1991). Figure 58.

Almost worldwide in temperate to subtropical climates. In North America, *Lemna minor* ranges north to a line between central Alaska, Great Slave Lake in the Northwest Territories, Hudson Bay in northern Manitoba, and the north shore of the Gulf of St Lawrence. In British Columbia, this species is found almost throughout the province; the large empty space on the map being probably a reflection of a shortage of collectors rather than an absence of the species. It floats on the surface of ponds and ditches, forming light green colonies of thousands of tiny discs; and is the most commonly seen duckweed in British Columbia.

Lemna minor L., in the broad sense, comprises a number of variants that have been distinguished as species. Two that are recorded for this province are *L. minor* L., in the strict sense, and *L. turionifera* Landolt. They may be distinguished as follows (Landolt 1975):

Key to Species in the *Lemna minor* complex

1a. Plant three- to five-veined, without purple pigment beneath; with one or two papillae (pimples) on the midline above, not producing rootless tubers. .... *Lemna minor sensu stricto*

1b. Plant three-veined in North America, purple beneath with anthocyanic pigment, with three to seven papillae along the midline above, producing rootless tubers. .... *L. turionifera*

Though the above distinctions appear to be clear in print, they are difficult to apply in practice. Pigmented and unpigmented plants may be found in one colony, sometimes even joined together in a vegetative mother-daughter relation. The papillae are hard to see except for the central one, the veins are often difficult to distinguish, and the overwintering tubers are produced only under appropriate conditions, and thus are not normally found. Much of our material cannot be clearly assigned to either species at sight.

Lemna turionifera occurs across North America and in eastern Asia and Hawaii. Identifiable specimens of it at the Royal British Columbia Museum are from Fort St John, Burns Lake, Vanderhoof, Nakusp, and Osoyoos, in British Columbia. Because of the uncertainly of distinguishing *L. turionifera* from *L. minor* in the majority of cases, records of both are included in the distribution map for *L. minor*.

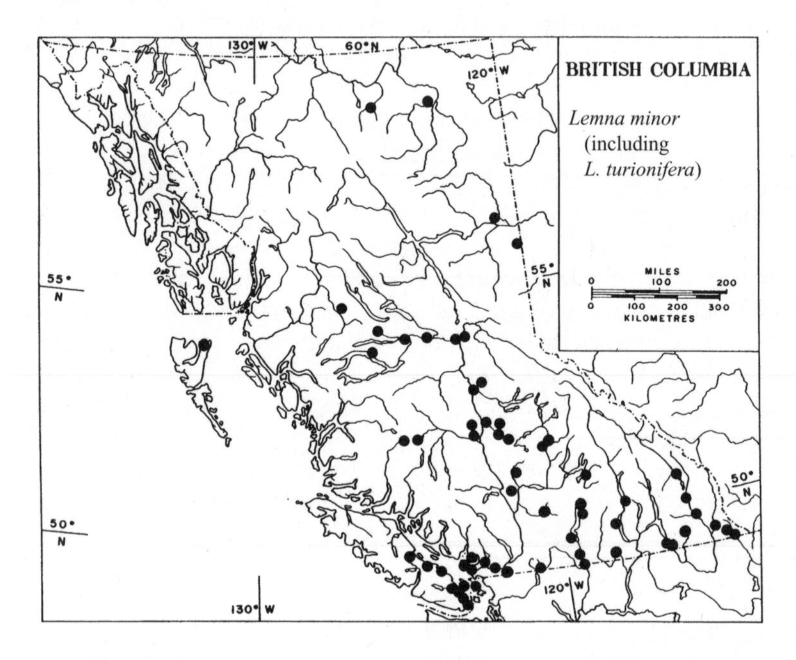

The Eurasian *Lemna gibba* L. (Inflated Duckweed), which differs from *L. minor* by its five-veined, hemispheric (inflated beneath) – rather than disclike – body, has been reported in the province (Hitchcock and Cronquist 1973). I have seen no currently accepted record of this species from British Columbia.

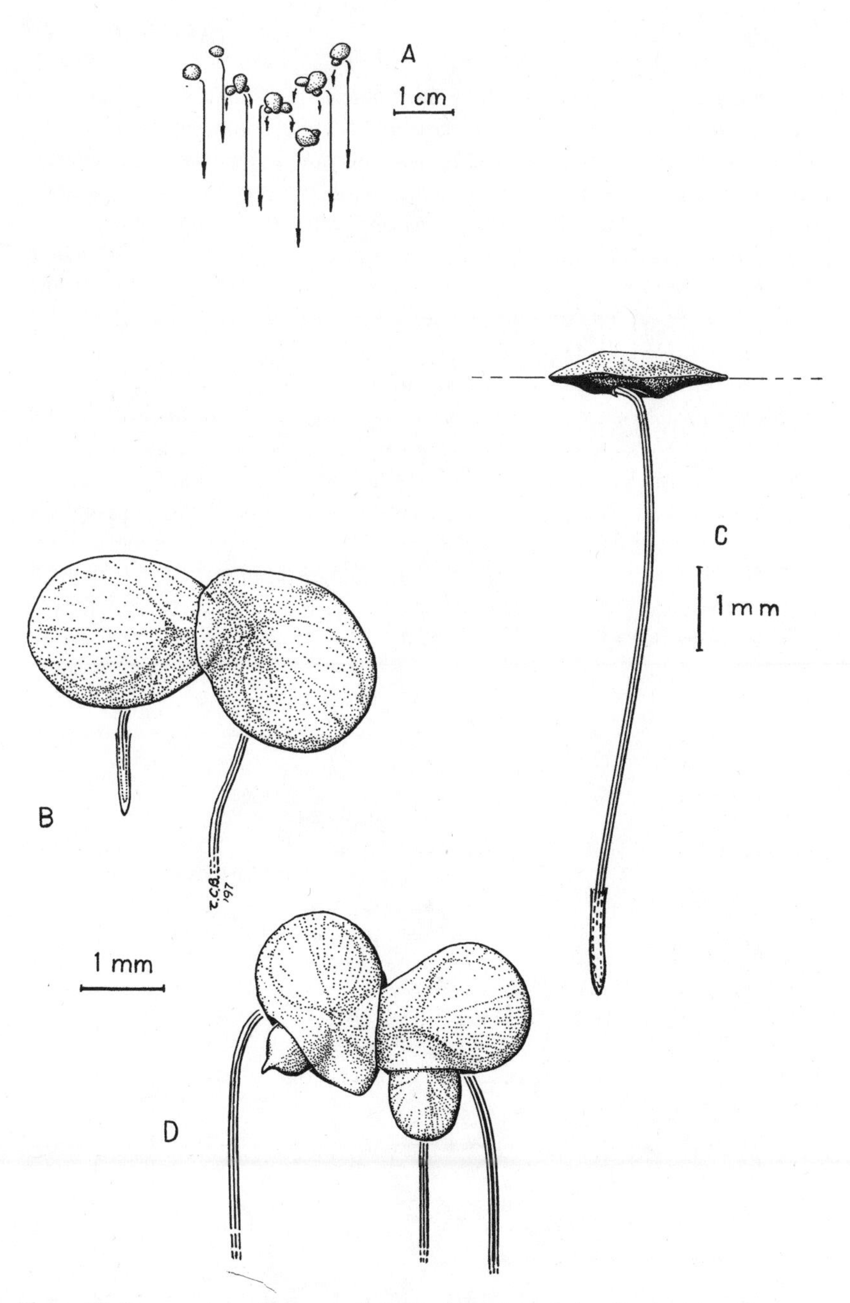

Figure 58. *Lemna minor*: A, group of plants, about natural size; B, plant, top view; C, young plant, side view; D, fruiting plant, top view.

Lemna trisulca L. **Star Duckweed**

Plant body rather elongate and lanceolate, tapering to a stalked base and acute apex, up to 13 mm long over all, dark green above and beneath, obscurely three-veined. Pockets on either side near the junction of the stalk with the main plant body give rise to young plants, which spread at right angles to the axis of the parent plant. Several "generations" of plants often remain attached to each other. At flowering, one flask-shaped ovary, tipped by a short style, is accompanied by two staminate flowers, of one stamen each, projecting from a pouch. 2n = 20, 40, 60, 80 (Gleason and Cronquist 1991). Figure 59.

Lemna trisulca is widespread in temperate to subtropical climates. In North America, it ranges from central Alaska to Labrador, reaching the Arctic coast in the western Northwest Territories, and from there southward through Canada and the United States. It is widespread across British Columbia.

Since it is usually submerged, this species is less commonly seen than *L. minor*. It is usually seen in still or slowly flowing, somewhat calcareous waters. This species may lie just beneath the water surface, forming looser-looking colonies than *L. minor,* or may lie on the bottom in water 0.5 to 1 metre deep, forming extensive mats. Plants float exposed on the surface only for rare episodes of flowering.

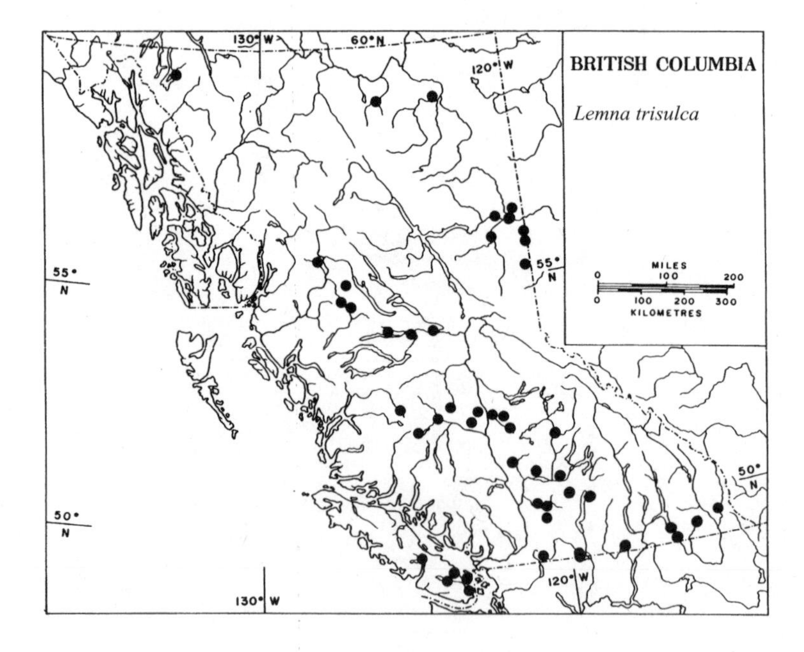

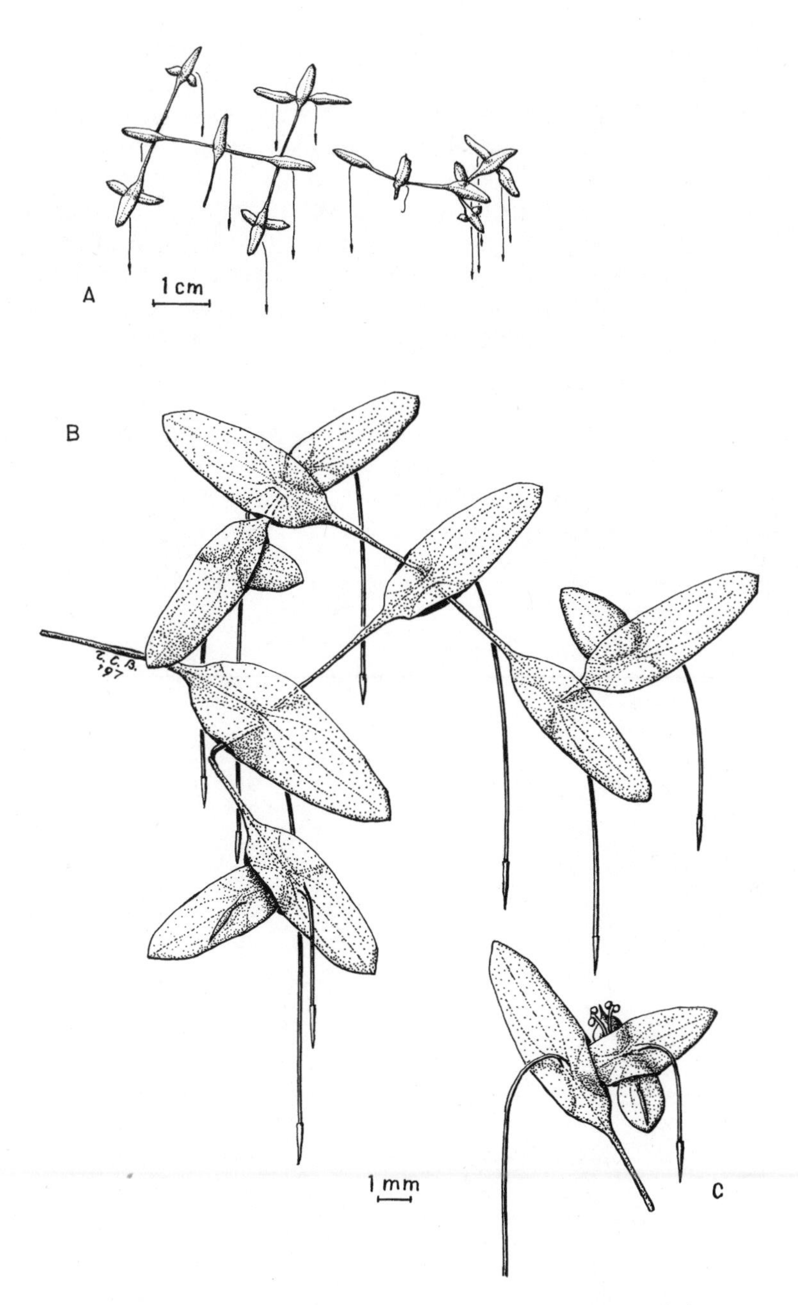

Figure 59. *Lemna trisulca*: A, small plant group, top view, about natural size; B, plant group, in top view, the lowermost plant overturned; C, flowering plant as seen from beneath.

The Genus *Spirodela* Schleiden

Plant body discoid, broadly obovate, radially 5- to 12-veined, green above, purplish beneath, with a cluster of rootlets centrally attached beneath, and axillary scales.

Young plants arise in succession from a pair of submarginal pockets near the base of the plant. The young plants are shortly stalked initially, but the stalk is soon lost. A degree of asymmetry has been reported for plants in flower, whereby one cleft (commonly the right-hand one as seen from above) is preferentially used for vegetative propagation, while the other gives rise to flowers.

One pistillate and two staminate flowers are produced together in a membranous spathe in one of the submarginal clefts. Anther with two chambers, opening longitudinally. Ovary with two ovules.

The genus *Spirodela* comprises three or four species, found on most continents, with one species here.

Spirodela polyrhiza (L.) Schleiden — Great Duckweed

***Lemna polyrhiza* L.**

Resembling a large *Lemna minor* plant as seen from above, the mature plant is 4 – 8 mm long by up to 6 mm wide, and dark purplish beneath. Rootlets 7 – 15, in a central tuft beneath.

The flowers are enclosed initially in a sacklike membranous spathe in a submarginal pouch covered by a small flap. 2n = 30, 40, 50 (Gleason and Cronquist 1991). Figure 60.

This species produces small tubers. These bodies are disc-shaped, 1 – 2 mm in diameter, veinless and commonly rootless, though they may develop a few short rootlets, and have one marginal reproductive pouch with a semicircular translucent flap. They are olive green in colour, darker than a normal plant, with reddish dots, but with little or no purple colour beneath. They contrast with the growing plants that sprout from them on short stems arising asymmetrically from the reproductive pouch. Tubers may be produced during the growing season. Floating tubers producing young, small plants, have been found in the Victoria area in early July.

Nearly cosmopolitan, in North America *Spirodela polyrhiza* is found from British Columbia to Quebec and Nova Scotia, and from there southward into tropical America. It is widespread in this province, but is less abundant than *Lemna minor*.

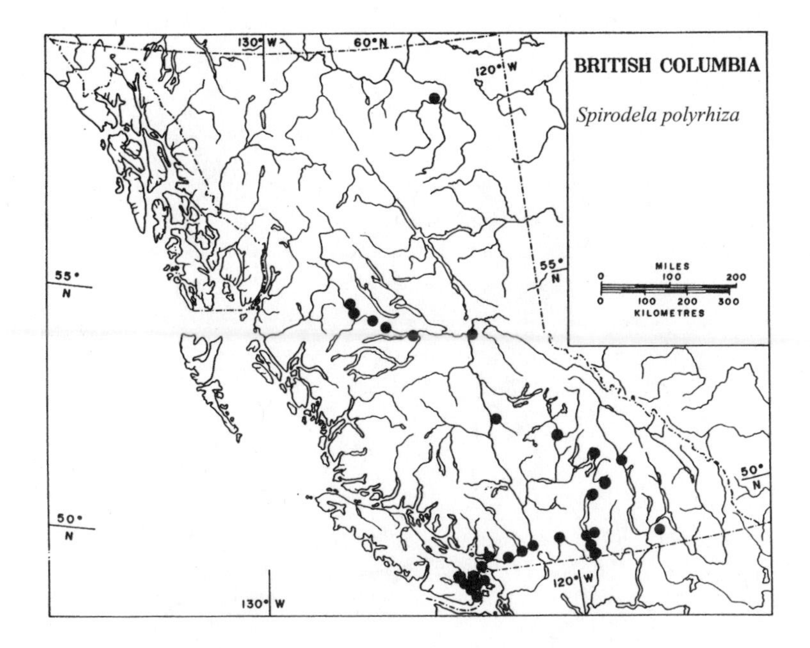

Found in the same habitats as *Lemna minor, Spirodela polyrhiza* reproduces much more sparingly, and seldom produces the large crowded colonies characteristic of that species. It commonly grows as isolated plants or small groups in *Lemna minor* colonies, where it may be distinguished by its conspicuously larger size.

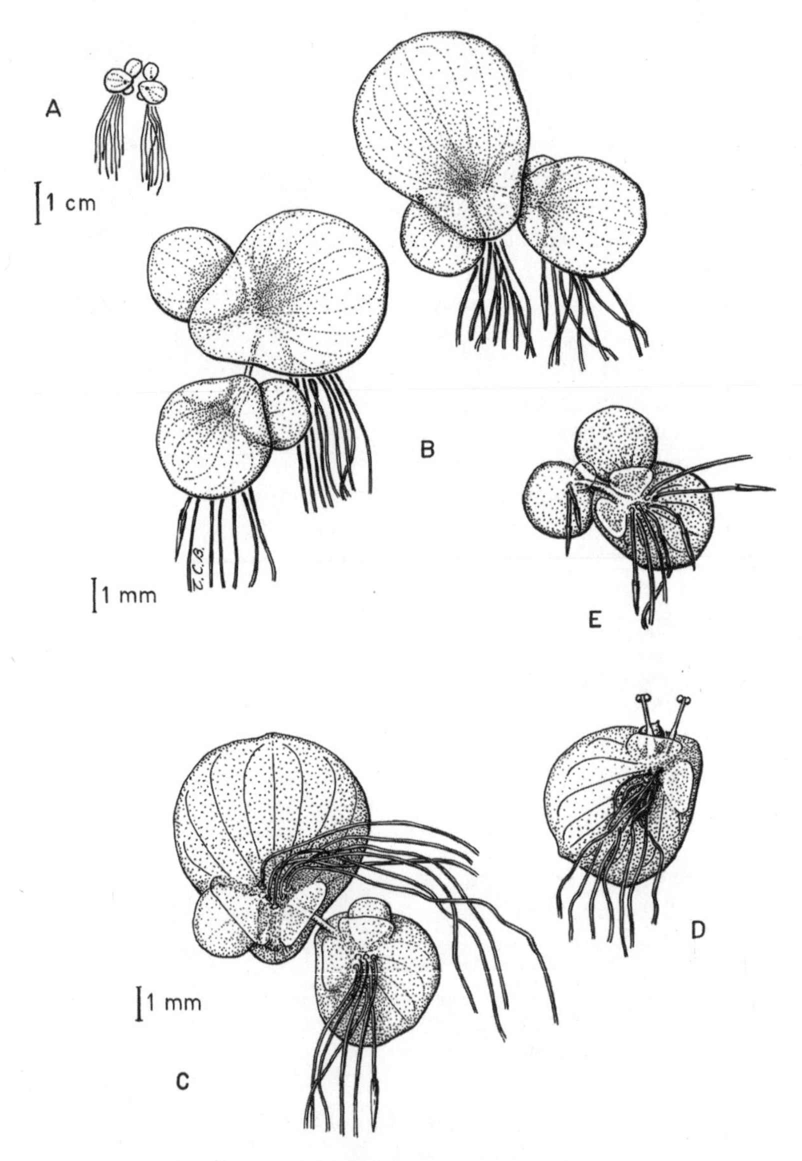

Figure 60. *Spirodela polyrhiza*: A, small plant group; B, plant group enlarged, top view; C, plants seen from beneath; D, flowering plant, seen from beneath; E, germinating tuber (on left) and first two generations of plants, seen from beneath.

The Genus *Wolffia* Horkel — Water-meal

Plant body hemispheric or ellipsoidal, without veins or roots or raphides, and with one basal reproductive pouch that produces young plants in succession.

Flowers in pairs on the upper surface – one staminate and one pistillate. The staminate flower is a one-chambered anther, and the pistillate flower is a globular ovary with a very short style and one ovule. Fruit a globular utricle, the seed with scanty endosperm.

A genus of six or seven species, mainly of tropical and subtropical distribution, *Wolffia* includes the smallest of all flowering plants.

Key to Species

1a. Plant (without young in pouch) ellipsoidal, a half to two-thirds as wide as its length..*W. borealis*

1b. Plant (without young) globose, almost as wide as its length. ....*W. columbiana*

Wolffia borealis **Water-meal**
(Engelmann *ex* Hegelmaier) Landolt and Wildi
***W. brasiliensis* Weddell var. *borealis* Engelmann *ex* Hegelmaier**
***W. punctata auct. amer. non* Grisebach**

Plant ellipsoidal, 0.5 – 1.2 mm long (average: 0.7 mm) and about a half to two-thirds as wide, often faintly dotted, with a pouch in the basal end from which young plants arise in succession (Dore 1957). 2n = 20, 30, 40. Figure 61.

Known from northeastern and western United States and southeastern Canada, this species scarcely enters British Columbia. It has been found in the lower Fraser River valley and on the Fraser River delta south of Vancouver, and at Creston (Ceska and Ceska 1980; Douglas et al. 1994).

At the few sites I have seen this species, it has formed a rich green "cream" on the surface of a pond or ditch, far out-reproducing the accompanying *Lemna minor*, which was scattered in only small groups within the *Wolffia* colonies. Owing to its thick (not flat) form, *Wolffia borealis* does not reflect the sky as does *Lemna minor*, and so appears a deeper, richer green colour when seen from above. It also has a slightly gritty feeling when handled in the water.

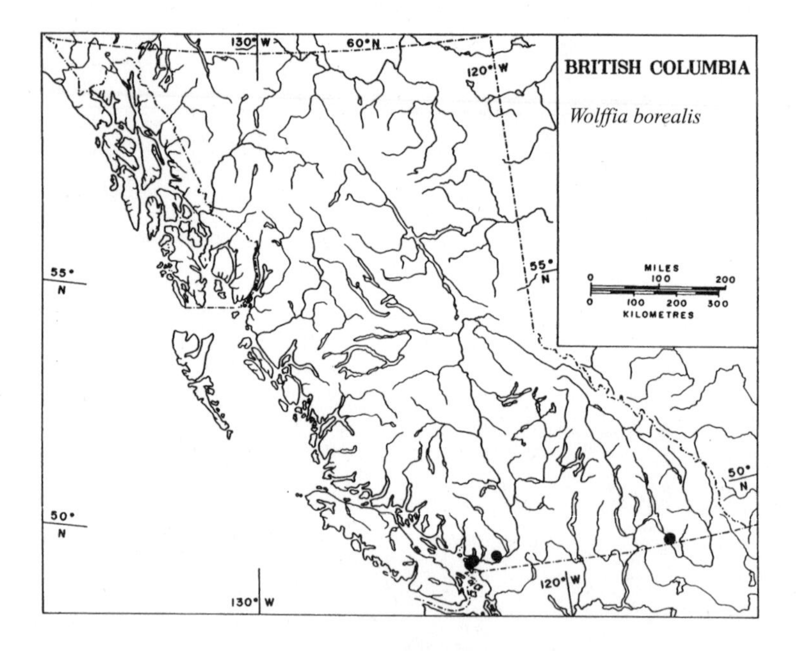

Figure 61 (facing page). *Wolffia borealis*: A, plant group, about natural size; B, plant, enlarged, in side view; C, plants, enlarged, seen from above.

Wolffia columbiana: D, plant group, about natural size; E, plants, enlarged, in side view; F, plants, enlarged, as seen from above.

Wolffia columbiana Karsten

Plant globose, 0.7 – 1.0 mm in diameter, almost as wide as its length, with the upper, barely exposed surface slightly flat. In life, the internal air spaces are discernible through the semitransparent epidermis. 2n = 30, 40, 50, 70. Figure 61.

Widespread in eastern and central North America, and also in South America, *Wolffia columbiana* is rare in western North America, though recorded in Oregon. In British Columbia, it has been found only at Saanich, near Victoria, where it is spreading on streams and lakes, apparently having reached this area in the 1980s (no map). The plant is easily overlooked unless present in very large numbers.

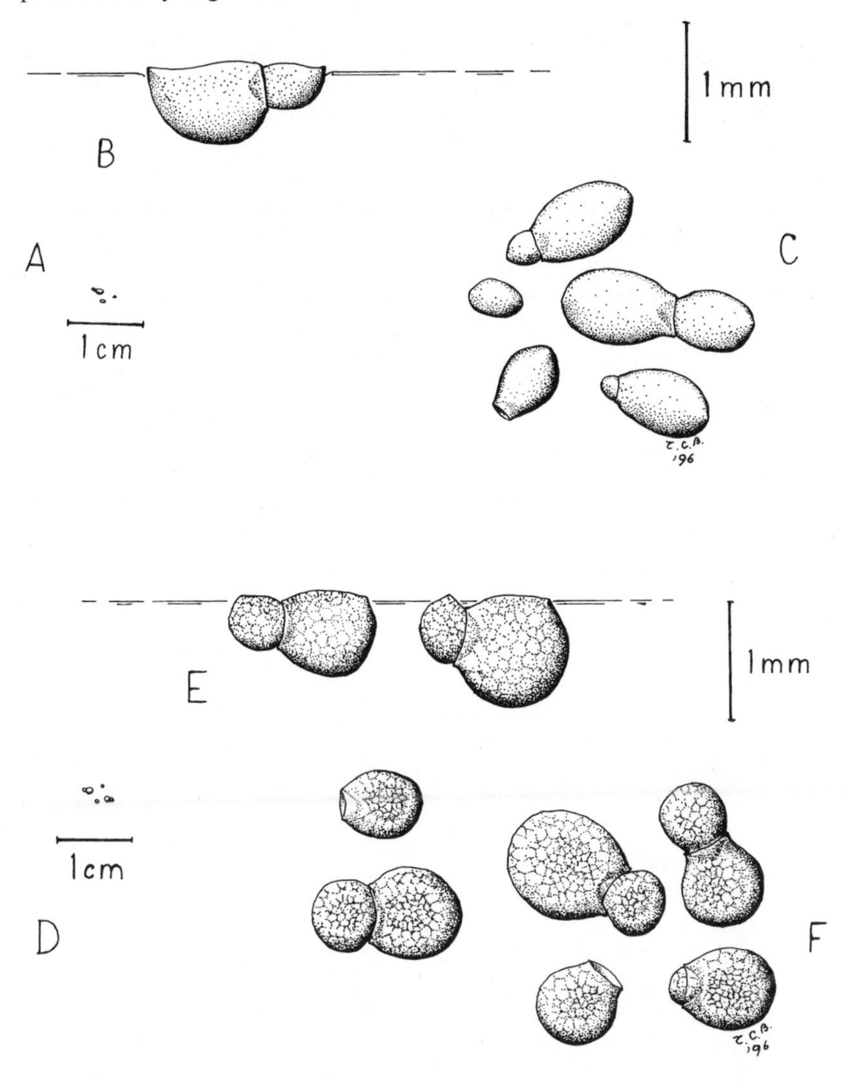

The Family Pontederiaceae — Pickerel-weed Family

Perennials of diverse form, with rhizomes or stolons, sheathing leaf bases and bracts, and coloured flowers in spikes, racemes or compact panicles, rarely single.

Flowers complete, the joined bases of the perianth parts and stamens forming a perianth tube or funnel. Sepals and petals coloured alike, often showy. Three or six stamens, often of different forms or sizes. Three carpels, connate but free from the perianth. Fruit a capsule or utricle. Seed with a slender straight embryo and abundant floury endosperm.

A small family comprising seven genera and 28 species, mostly tropical.

Key to Genera

1a. Erect, with cordate or oblong leaves. Flowers in spikes, bilaterally symmetric, blue in ours. Six stamens per flower, unequal in length. ....*Pontederia*

1b. Submersed, trailing or floating, with linear to lanceolate, ribbonlike leaves. Flowers solitary, radially symmetric, pale yellowish. Three stamens per flower, all alike..*Zosterella*

The Genus *Pontederia* L. — Pickerel-weed

Stout, emergent perennials with thick rhizomes and erect leaves and flowering shoots. Leaves ovate, cordate-based, the basal ones on long petioles, the stem leaves short-petioled from dilated sheathing bases.

Inflorescence a compound spike, the flowers opening successively in each lateral spikelet. Perianth tube funnel-shaped; it and the perianth lobes minutely villous on the outer surfaces. The upper three lobes coherent into a lip. Six stamens, unequal in length; the anthers attached at their midpoints. Ovary three-loculed, but only one locule fertile. Fruit a one-seeded utricle enclosed in the distended perianth tube.

There are three species, of temperate to tropical Americas. One species has been found in British Columbia.

Pontederia cordata L. **Pickerel-weed**

Leaves with stout, erect petioles and cordate-ovate, pinnately parallel-veined blades with rounded tips. Inflorescence dense, showy, the blue flowers puberulent on the outside, and with one or two yellow spots within on the upper lip of the perianth. Stamens very unequal: three projecting well out of the perianth tube, and three very short, with their anthers in the mouth of the perianth tube. Anthers blue, rather heart-shaped. Fruit with six longitudinal toothed ribs. 2n = 16. Figure 62.

Widespread in eastern North America. Introduced near Port Alberni, but not reported recently; one small clump seen at Glen Lake, near Victoria, appears to have been planted (no map). Within its natural range, this species grows in still lake water to a metre deep, and may form extensive colonies along the margins of shallow lakes.

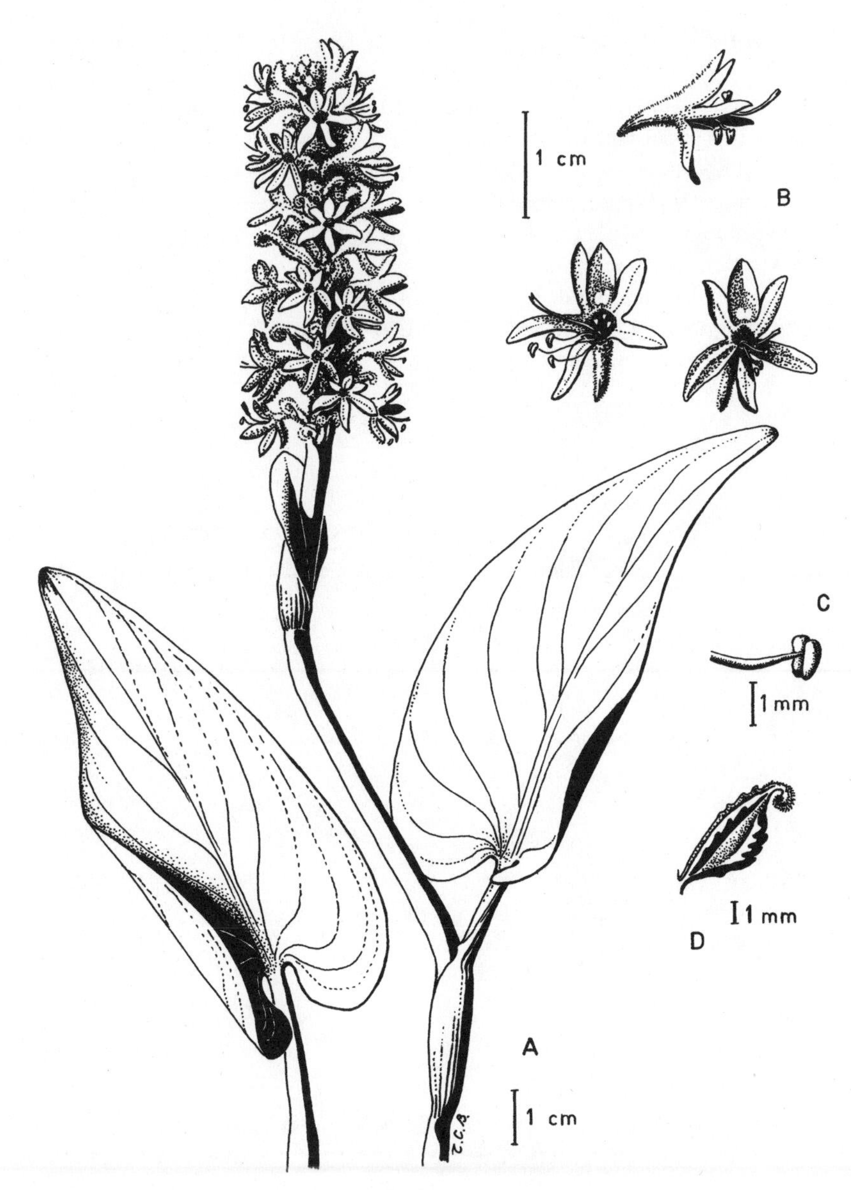

Figure 62. *Pontederia cordata*: A, upper part of flowering shoot, and leaves; B, flowers; C, stamen; D, maturing fruit.

The Genus *Zosterella* Small — Water Star-grass

Submersed to floating plants with slender stems. Leaves linear, ribbonlike, sessile. Flower solitary in a spathe, with a slender, sometimes elongate perianth tube and three each of equal, spreading, pale yellow, linear sepals and lanceolate petals. Three stamens, all alike, with somewhat inflated filaments and sagittate anthers that coil up after shedding their pollen. Fruit a capsule with few to many longitudinally ribbed seeds.

One species, often united with *Heteranthera*, as in the first edition of this book, but now usually segregated on account of its similar stamens and its linear, ribbonlike, mostly submerged leaves.

Figure 63. *Zosterella dubia*: A, flowering plant; B, ligular sheath; C, flower; D, seeds.

Zosterella dubia (Jacquin) Small — Water Star-grass
***Heteranthera dubia* (Jacquin) MacMillan**

Rather grasslike plant, commonly submersed except for the open flowers. Stem terete or rather flat.

Leaves linear to narrowly lanceolate, tapering to base and acute apex, sessile, the base adnate to the lower half of a tubular sheath. Sheath with a conspicuous ligular extension that is a first rounded at the tip, but later breaks down, leaving two persistent fibres as slender appendages. Leaves often subopposite where a flower arises.

Flower solitary, emerging from an axillary spathe. Perianth tube slender, 2 – 7 cm long. Sepals and petals pale yellow, spreading, the sepals linear, and the petals lanceolate. Three stamens, alike, with short flat or dilated filaments, and basally attached, sagittate-based anthers. Capsule unilocular, with several longitudinally ribbed seeds. Flowers often do not open, but are self-pollinating in bud. $2n = 30$. Figure 63.

Zosterella dubia is known from the southern parts of British Columbia, Ontario and Quebec, but not the Prairie provinces, and ranges southward through the United States to tropical America. In British Columbia it is found in the southern interior, northward to Golden and to Shuswap Lake, and westward to Hope. A collection from near Holberg, on northwestern Vancouver Island, would appear to represent a recent introduction there. It is found in shallow (to a metre deep) water of lakes, ponds or slow, sometimes muddy, streams.

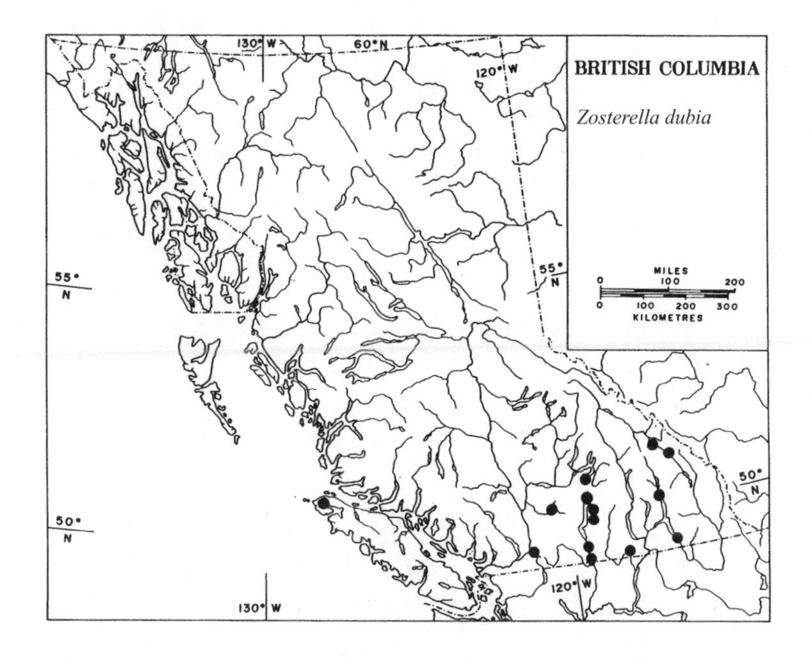

The Family Typhaceae — Cat-tail Family

Rhizomatous perennials with erect, straplike, basally sheathing leaves and unbranched flowering stems. Tracheae present in conducting strands throughout the plant.

Flowers in very dense terminal and subterminal spikes, the pistillate and staminate flowers in distinct segments of the spikes, with the staminate flowers in the apical segment. Bracts of the inflorescence (one for each segment) usually transitory, dying and disintegrating early. Bracts of the flowers minute, hidden among the flowers, or absent.

Staminate flowers have two to four stamens, the filaments distinct or united. Pollen grains separate or coherent in groups of four (tetrads) when released. Pistillate flower with a perianth reduced to a tuft of hairs about as long as the pistil, and a unicarpellate ovary on a stipe and tipped by a slender style and stigma, and containing one ovule. Sterile, stigma-less flowers are often mixed among the fertile ones.

The fruit is an achenelike, one-seeded follicle, which is released by the breaking away of the pistillate flower from the axis of the spike. This fruit is light, and buoyant by reason of the tuft of perianth hairs; and, with its contained seed, is borne away on the wind. The subsequently released seed contains an embryo embedded in endosperm made up of starch, oil and protein.

The family consists of one genus, *Typha*, with 10 – 15 species, of cosmopolitan distribution.

The Typhaceae and Sparganiaceae are obviously closely related, and some botanists (Thorne 1976) treat them as one family, the Typhaceae. Their relationships to other monocotyledons however, are more distant and obscure, and the subject of some disagreement; their combination with the Arales in a suborder Ariflorae (Thorne 1976) being but one of a number of postulated assignments for them. Formerly included in the order Pandanales, they have been separated by more recent workers from that order, which otherwise is characterized by multicarpelled, multi-ovuled ovaries and an arborescent or vinelike habit of growth.

The Genus *Typha* — Cat-tails

Key to Species

1a. Pistillate and staminate segments of the spike contiguous. Stigmas ovate to lanceolate. .... *T. latifolia*

1b. Pistillate and staminate segments separated by a bare stem segment 2

2a. Pistillate flowers with minute bracts. Stigmas filiform. Pollen grains shed as separate single grains. .... *T. angustifolia*

2b. Pistillate flowers without bracts, or stigmas lanceolate to ovate, or pollen grains shed in tetrads (cohering in groups of four)...*T.* x *glauca*

Typha angustifolia L. **Narrow-leafed Cat-tail**

Similar to *T. latifolia*, but with narrower leaves 3 – 10 cm wide, and the spike segments separated by a bare segment of the stem. Spikes slender, the pistillate segment minutely bristly in surface view because of the filiform stigmas, with inconspicuous bracts among the flowers. The pollen grains are shed as separate grains. 2n = 30. Figure 64.

Typha angustifolia is widespread across the northern hemisphere, but with a narrower latitudinal range than *T. latifolia*. In North America, it is found from southern Manitoba, southern Ontario, Quebec and Nova Scotia southward to California and Nebraska (Scoggan 1978). It has been reported from the Vancouver area since the first edition of this book (no map), but I have not seen that material. Other records that I have examined were either *T. latifolia* or *T.* x *glauca*. I drew the illustration from Quebec material.

Typha x *glauca* Godron (= *T. angustifolia* x *latifolia*)

***T. latifolia* forma *ambigua* (Sonder) Kronfeld**

This population, combining in diverse ways the characteristics of *T. angustifolia* and *T. latifolia*, is thought to be of hybrid origin, and as such it is identified by the above binomial and formula, the "x" denoting its hybrid status. Plants of this nature have been seen at widely scattered points in British Columbia. I recall a sizeable stand on the Kootenay River below Creston. Initially diagnosed as *T. angustifolia*, it subsequently proved to have its pollen grains in tetrads; thus demonstrating its mixed origin. Since their seeds may be carried long distances by wind, and their rhizome fragments can be carried away by water currents to become established far from their points of origin, members of this hybrid population may appear in locations where one parent or the other is absent. (No illustration or map.)

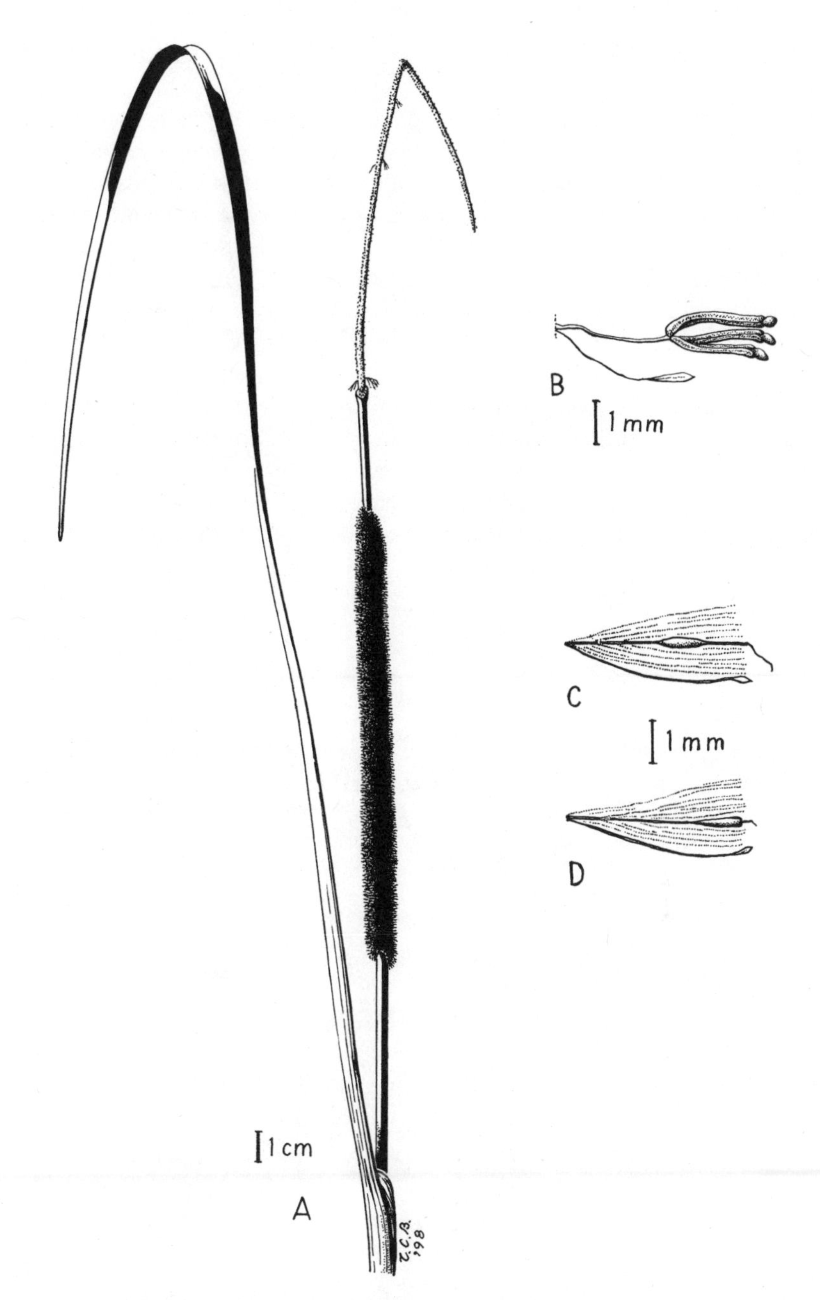

Figure 64. *Typha angustifolia*: A, fruiting spike and leaf; B, staminate flower; C, fertile pistillate flower and bract; D, sterile pistillate flower and bract.

Typha latifolia L. — Common Cat-tail

Robust plant, 1 – 3 metres tall, with a stout rhizome, the two-ranked leaves greyish green, 8 – 20 mm wide, with inconspicuously sheathing bases.

Inflorescence with staminate and pistillate segments of the spike normally contiguous, but occasionally separated by a bare stem segment. Pistillate flowers without bracts, with ovate to lanceolate stigmas. Sterile pistillate flowers club-shaped, rounded or terminated by a short, acute tip, and with the short, aborted style above. Fruiting spike 1 – 3 cm thick, dark brown at maturity. Pollen grains released in tetrads, and borne away on the wind. 2n = 30. Figure 65.

Typha latifolia is widespread in the northern hemisphere. In North America it reaches northward to central Alaska, near Great Bear Lake in the western Northwest Territories, James Bay and Newfoundland, and ranges from there southward to Mexico. It is common in many parts of this province. It is a characteristic dominant of many shallow-water marshes. Multiplying and spreading by its rhizomes, it commonly forms extensive, dense colonies, almost to the exclusion of other species.

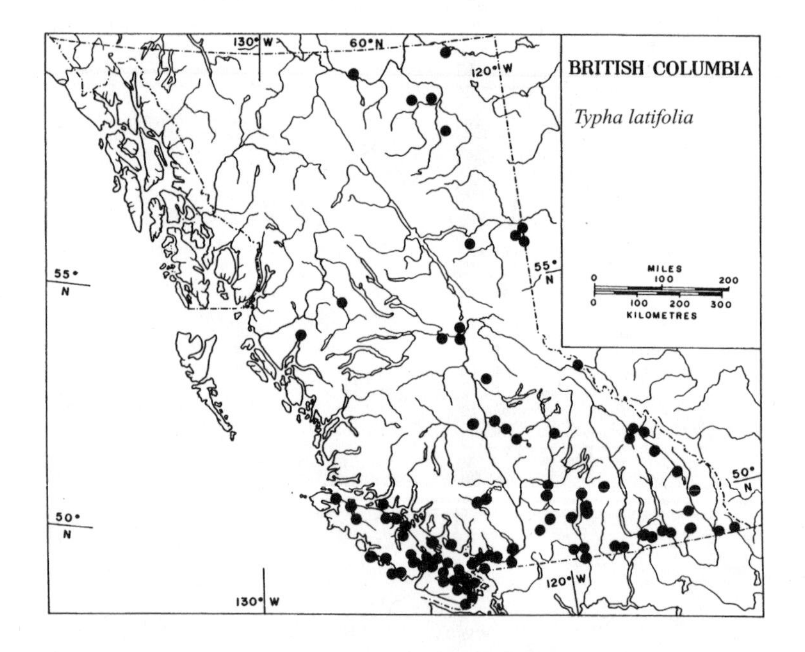

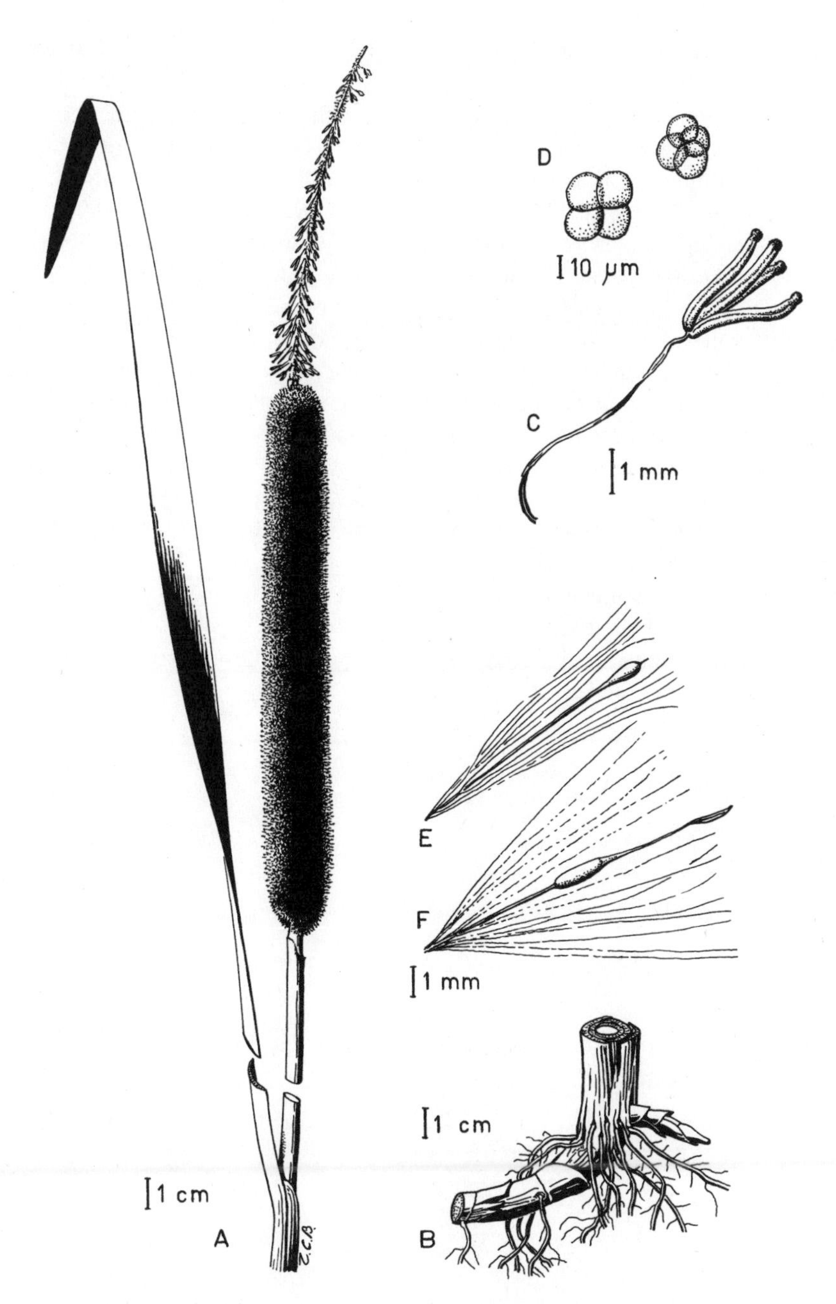

Figure 65. *Typha latifolia*: A, inflorescence at fruiting stage, and leaf; B, rhizome segment and base of shoot; C, staminate flower; D, pollen tetrads; E, sterile pistillate flower; F, fertile pistillate flower at fruiting stage.

The Family Sparganiaceae — Bur-reed Family

Aquatic, monoicous, rhizomatous perennials with alternate, linear leaves with sheathing bases, and tracheae in the conductive tissues throughout the plant.

Staminate and pistillate flowers in separate globose heads, the pistillate heads below, and the staminate heads above, in simple or branched inflorescences; the pistillate heads at least, subtended by conspicuous, leaflike bracts. In some species, some of the pistillate heads appear to be supra-axillary through fusion of axillary peduncles with the main axis of the inflorescence.

Staminate flowers are, individually, almost indistinguishable; three to five stamens are surrounded by or mixed with several narrow, scale-like tepals. Pistillate flowers short-pedicelled in most species, but sessile in *Sparganium eurycarpum*; they have up to six scale-like tepals and an ovary of one or two carpels, and bear a beaklike style and one or two stigmas. Locules one or two, each containing one ovule. The fruit is an achene tipped by the stylar beak (often including the persistent stigma) and shed with the tepals attached. The seed has an embryo embedded in starchy endosperm.

This family is assumed to be wind pollinated, but bees have been seen visiting the flowers of some species.

This family consists of the one genus *Sparganium*, distributed around the world in temperate and cooler climatic zones.

The Genus *Sparganium* **Bur-reeds**

Characters of the family. Some sections of the genus have proven difficult to classify; and the number of species described varies with varying treatment. The genus has around 20 species.

Sparganium species are very plastic in their character expressions. Superimposed on the range of genetically induced variation is variation in response to environmental factors such as altitude and water depth, and some apparently random variation for which no obvious cause can be distinguished. In some species complexes, this plasticity, whatever its origin, is a cause of much difficulty in identifying specimens; and the classification and nomenclature of the species is surprisingly difficult for a relatively small genus.

Since they reproduce vegetatively by spreading rhizomes as well as by seeds, these species commonly form extensive colonies of "plants" that are in fact merely branches from a single original plant. Such a colony is a clone, in which all the "plants" are genetically identical. Even so, a surprising range of variation may be seen within a colony in such characters as leaf width and stiffness, and peduncle adhesion.

Key to Species

1a. Stigmas normally two per flower, at least on some flowers, 2 mm or more long. Fruit sessile and pear-shaped with the wide end distal. Inflorescence branched, with many staminate heads, and pistillate heads confined to axillary branches. Anthers obovoid. Leaves stout, erect and arching, 8 – 20 mm wide. ....*S. eurycarpum*

1b. Stigma single per flower, 2 mm or less long. Fruit tapering at both ends, stipitate. Inflorescence often unbranched. Pistillate heads on main axis. Anthers oblong or ellipsoidal. Leaves 1 – 10 (– 15) mm wide. 2

2a. Staminate head usually one, rarely two. Achene beak 1.7 mm or less long. Stigma 0.8 mm or less long. Leaves usually floating. 3

2b. Staminate heads two to several. Achene beak over 1.5 mm long. Leaves often ascending and emergent. 5

3a. All pistillate heads axillary and usually sessile. Achene beak 0.5 – 1.5 mm long. Stigma 0.3 – 0.8 mm long. Leaves 2 – 8 mm wide, thin and translucent. ....*S. natans*

3b. At least some pistillate heads supra-axillary. Lowest pistillate head usually peduncled. Leaves thick and opaque. 4

4a. Achenes 3.5 – 4.5 mm long, beakless or with a short conical beak up to 0.5 mm long. Stigma 0.2 – 0.5 mm long. Heads usually clearly separate. Leaves 1 – 5 mm wide. ....*S. hyperboreum*

4b. Achenes 5 – 6.5 mm long, with a conical beak 1 – 2 mm long. Stigma 0.6 – 0.7 mm long. All but the lowest heads aggregated. ....*S. glomeratum*

5a. Pistillate heads all axillary or on axillary branches. Stigma 0.8 – 2 mm long. .... *S. americanum*

5b. At least some pistillate heads supra-axillary 6

6a. Achene beak distinctly curved and flat. Stigma ovate, 0.4 – 0.7 mm long. Anthers 0.8 mm or less long. Inflorescence commonly branched. Leaves floating, or some emergent..................... *S. fluctuans*

6b. Achene beak straight or slightly curved, terete. Stigma linear, 1 – 2 mm long. Anthers 0.8 – 1.5 mm long. ....*S. angustifolium* complex

Figure 66. *Sparganium americanum*: A, upper part of fruiting plant; B, achene, with attached tepals; C, stamen.

Sparganium americanum Nuttall American Bur-reed

Slender or stout, usually erect and emergent plant. Leaves three-angled, ascending and emergent, or sometimes ribbonlike and floating, 4 – 20 mm wide.

Inflorescence simple or branching. All pistillate heads, or branches bearing them, axillary. Main axis with one to five pistillate heads and five to nine staminate heads. Branches, when present, have fewer heads (Fassett 1957). Anthers 0.8 – 1.2 mm long. Stigmas 1 – 2 mm long. Fruiting heads 1.5 – 2.5 cm in diameter. Achene tapering to ends, dull or slightly lustrous, with a beak 1.5 – 5 mm long, including the 0.8 – 2 mm stigma. Figure 66.

Widespread in eastern North America, in shallow water and on shores, this close relative of the *S. angustifolium* complex is retained here because of reported occurrences in British Columbia by Fernald (1950), Boivin (1967), Taylor and MacBryde (1977) and Scoggan (1978). The Royal British Columbia Museum has one specimen, collected in 1980 on Lulu Island in the Fraser River delta (Greater Vancouver; no map), that has all its pistillate heads, or their peduncles, strictly axillary. In 1981, Vernon L. Harms annotated this Lulu Island material as *S. chlorocarpum* Rydberg forma *chlorocarpum*. But I feel that this material is just as likely to be *S. americanum*. More collections from this Lulu Island population are desirable. Of three collections from White Rock, alluded to in the first edition of this book, each has one head very shortly supra-axillary, with the adhesion shorter than the radius of the head and thus difficult to see; the other heads being axillary. The stigmas of all these plants are about 2 mm long. I now regard the White Rock material as *S. angustifolium* ssp. *emersum* var. *multipedunculatum*.

In view of the above argument, I have retained this species in this edition, although at present, I know of no other material from British Columbia, that can be assigned to *S. americanum*. The illustration (figure 66) in this edition is based on material of *S. americanum* from eastern Canada.

Sparganium angustifolium Michaux

S. simplex* Hudson, *nomen illegitimum

Including:

- ***S. simplex* var. *angustifolium* (Michaux) Torrey**
- ***S. simplex*. var *multipedunculatum* Morong (*S. multipedunculatum* (Morong) Rydberg) = (*S. angustifolium* var. *multipedunculatum* (Morong) Brayshaw)**
- ***S. emersum* Rehmann = *S. angustifolium* ssp. *emersum* (Rehmann) Brayshaw**
- ***S. chlorocarpum* Rydberg = *S. angustifolium* var. *chlorocarpum* (Rydberg) Brayshaw**

Sparganium angustifolium is a highly variable plant with creeping buried rhizomes. Stems either slender and submerged, floating except for an emergent inflorescence, or erect and emergent.

Leaves arising from the plant base in spring commonly floating and ribbon-like, up to a metre long. Later leaves, arising from the flowering stem, may be floating, 2 – 5 (– 8) mm wide, convex dorsally, and dilated at the sheathing bases, or stiffly emergent, 20 – 50 cm long and up to 15 mm wide.

Inflorescence usually unbranched, the three or four pistillate heads normally peduncled except the uppermost, and at least one of them supra-axillary, its peduncle partially joined to the axis of the inflorescence. Staminate heads up to eight, the stamens white at flowering time. Anthers 0.8 – 1.5 mm long by 0.5 mm wide.

Fruiting heads 1 – 3 cm in diameter, greenish or brown. Achenes tapering to ends, sometimes with a median constriction; the body 3 – 5 mm long, on a stipe up to 2.5 mm long, and with a straight to slightly curved beak 2 – 5 mm long, including the 1 – 2 mm long stigma. Four to six tepals on the fruit, spatulate, fringed above, arising from the middle of the stipe and reaching about half way up the achene. 2n = 30. Figures 67 and 68.

Key to Subspecies and Varieties

1a. All leaves normally floating even on flowering shoots, and 5 mm or less wide, seldom wider, not translucent-edged. Stems normally submerged or floating, except for the inflorescence. Bracts dilated at base. Achene beak about 2 mm long. Stigma about 1 mm long. ....ssp. *angustifolium*

1b. Spring leaves commonly floating, but flowering stems and their leaves normally upstanding out of the water, the leaves usually over 5 mm wide, often translucent-edged at base, and triangular in cross-section. Achene beak 2 – 5 mm long. Stigma 1.5 – 2 mm long. ....ssp. *emersum*:2

2a. Bracts scarcely dilated at base. Leaves 3 – 8 mm wide. ....var. *chlorocarpum*

2b. Bracts slightly to strongly dilated at base. Leaves 5 – 15 mm wide....3

3a. Basal leaves strongly keeled at base: Y-sectioned. Mature fruit 12 – 15 mm long...var. *emersum*

3b. Basal leaves not strongly keeled at base: V-sectioned. Mature fruit 10 – 12 mm long. ....var. *multipedunculatum*

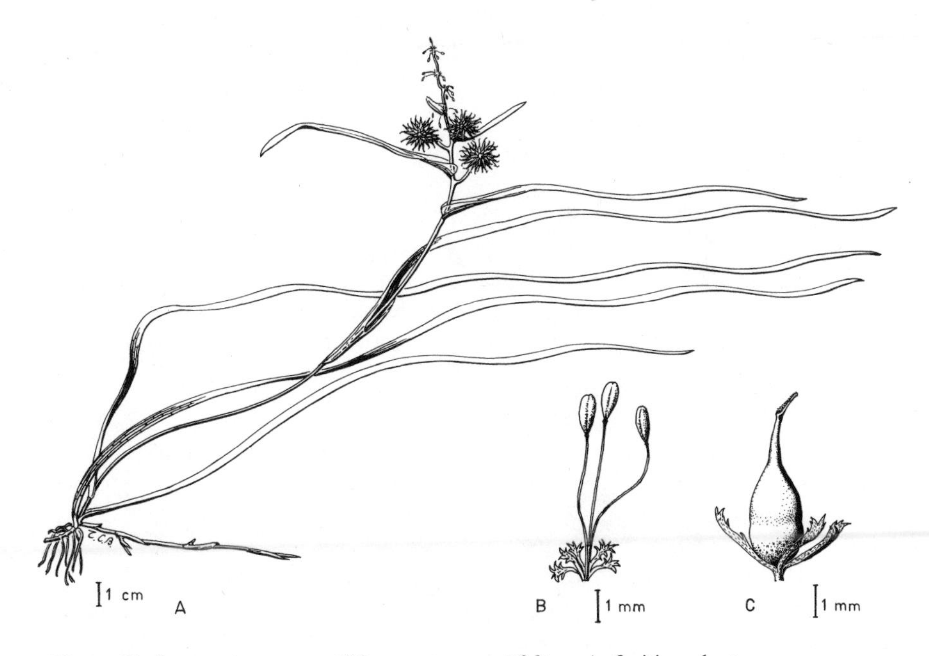

Figure 67. *Sparganium angustifolium* ssp. *angustifolium*: A, fruiting plant; B, staminate flower; C, fruit (achene with attached tepals).

Treatments of this complex vary. Some or all of these subspecies and varieties are treated as species by some authors. Typical examples of these entities are distinct and recognizable; but since they appear to intergrade in character, and interbreed freely, the conservative approach used here appears to be the most usable in practical application.

Both subspecies are circumboreal in range. In North America they occur in one variety or another over most of the continent, reaching south to California and north to the Arctic coast and Greenland.

Cook (1980) reports that, in Europe, hybrids between these subspecies (treated there as species) are abundant, as are hybrids between this complex and *S. gramineum* Georgi, the European relative of the North American *S. fluctuans*.

Subspecies *angustifolium* is found typically in cold ponds and ditches, with silt or muck bottoms, from the shore to water depths of a metre or more.

Sparganium angustifolium is quite variable in form. On Vancouver Island, plants appear more slender than usual. Especially at subalpine levels in the central part of the Island, plants may have almost filiform leaves only 1 – 2 mm wide, and small heads only about a centimetre in diameter; and thus tend to resemble *S. hyperboreum* in gross appearance. The number of staminate heads, and the longer beaks and stigmas than are typical of *S. hyperboreum*, indicate that these plants are *S. angustifolium*.

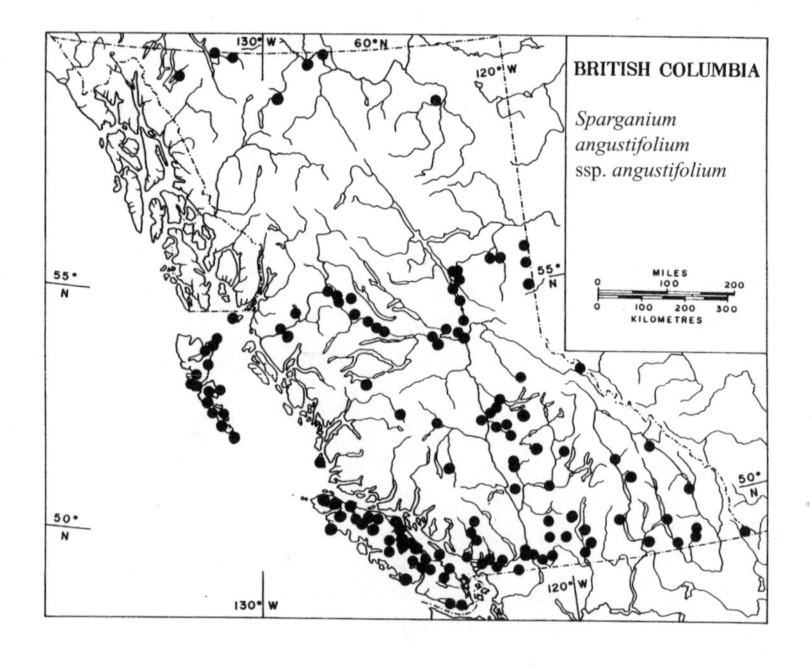

Sparganium angustifolium Michaux ssp. *emersum* (Rehmann) Brayshaw

S. erectum L. in part
S. simplex Hudson, _nomen illegitimum_ (Reveal 1970)
S. emersum Rehmann

The distinction between ssp. *angustifolium* and ssp. *emersum*, while appearing clear when typical plants of these entities are compared, are rendered very indefinite when large populations are examined, because of the frequent occurrences of individuals of ambiguous identity, which raise questions as to their true positions in this complex. I further suspect that ecological conditions may be involved, such that ssp. *angustifolium* may develop as a deep-water ecotype of ssp. *emersum*. Hence the most practical treatment appears to be to combine these two 'species" as subspecies of *S. angustifolium*, whose name has priority of publication.

Note on the Nomenclature

Regarding the use of *S. erectum* L., Linnaeus (1753, p. 971) included in his *Sparganium erectum* plants both with branched and unbranched inflorescences. He indicated that he considered the branched variety to be the typical one by citing under it *Sparganium ramosum* Bauhin, and by following that with the variety ß *Sparganium non ramosum* Bauhin. Examination of photographs (Savage 1945) of specimens in the Linnaean Herbarium reveals only

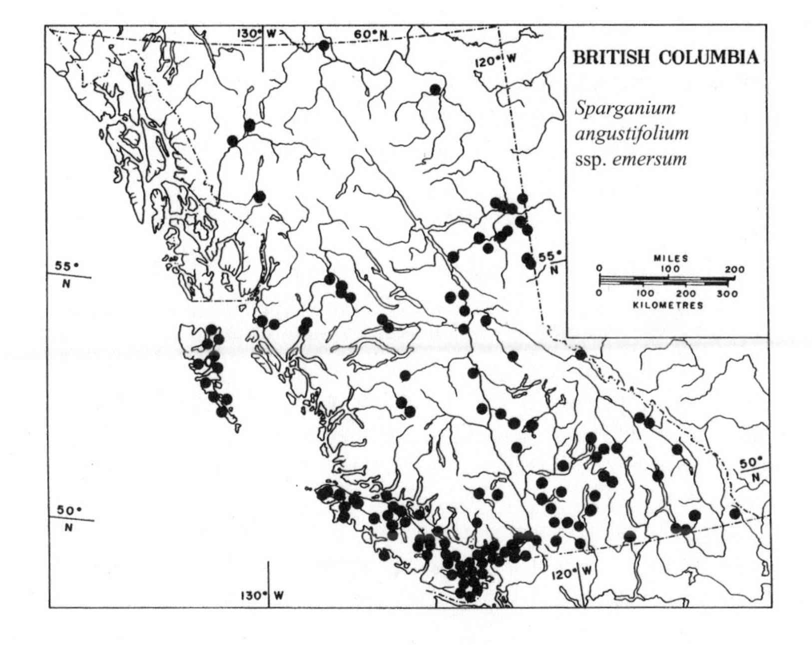

one specimen (#1095.1) identified as *S. erectum*. This specimen resembles *S. angustifolium* ssp. *emersum*, being unbranched, with two of its pistillate heads supra-axillary and erect leaves triangular in cross-section. Another specimen (#1095.3) resembles *S. angustifolium* ssp. *angustifolium*, as we today understand that entity. It bears no identification on the front of the sheet, but a note on the reverse states: "*Sparg: non ramosum*…". There is thus a discrepancy between the text and specimens. See also the note under *S. eurycarpum*.

Sparganium angustifolium ssp. *emersum* var. *chlorocarpum* (Rydberg) Brayshaw

***S. chlorocarpum* Rydberg**

This entity was distinguished originally by Rydberg from *S. simplex* on the leaf characters mentioned in the above key, but whose expressions may be partly modified by water depth and other considerations, and on characters of the fruit (10 – 12 mm long, green, dull, with a stigma about 1.5 mm long, as opposed to fruit 12 – 15 mm long, brownish, shiny, with a stigma about 2 mm long in *S. simplex*) that may at least partly reflect the stage of maturity of the fruit. These differences, while often discernible, do not always correlate with each other. It is felt that the overall differences are not adequate to distinguish entities at the rank of a full species, but rather to define varieties. Further, if the circumscription of *S. simplex* is taken as broad enough to include var. *multipedunculatum* (with fruit 10 – 12 mm long), as originally done by Morong, it is broad enough to include var. *chlorocarpum*.

Sparganium angustifolium ssp. *emersum* var. *multipedunculatum* (Morong) Brayshaw

***S. simplex* Hudson var. *multipedunculatum* Morong**
***S. multipedunculatum* (Morong) Rydberg**
***S. emersum* Rehmann var. *multipedunculatum* (Morong) Reveal**

This variety, like var. *chlorocarpum*, has a transcontinental range in North America. In *Sparganium angustifolium* and its relatives, the degree of adhesion between the peduncles of the pistillate heads and the main axis of the inflorescence may vary widely, and when large populations are examined, some individuals are found in which the adhesion is shorter than the radius of the head, and is thus concealed. These individuals, if looked at in isolation from their natural companions, can be taken as *S. americanum* Nuttall. They should be compared, if possible, with other members of the population to see if the concealed short adhesion occurs systematically throughout the population, or if it represents only the limit in one direction, in one individual, of a character that is more normally expressed in other members of the population.

Figure 68. *Sparganium angustifolium* ssp. *emersum*: A, upper part of plant and fruiting inflorescence; B, fruit (achene with attached tepals).

Sparganium eurycarpum Engelmann in Gray — Big-headed Bur-reed

***S. erectum* L., in part**
***S. greenei* Morong = S. *eurycarpum* var. *greenei* (Morong) Graebner**

This species is larger and coarser than our other species of *Sparganium*. It is rhizomatous and colonial, with shoots that are always emergent and upright in habit. Stems are coarse, up to 1.2 (or rarely to 2) metres tall, commonly over-topped by the leaves, and branching from the upper axils. Leaves are 1 – 2 cm wide, broadly triangular in cross-section, and keeled except toward their tips.

Inflorescence large, branching, with few pistillate and many staminate heads. The green pistillate heads are confined to low positions on the lateral branches and are often absent from the main axis; all heads appearing sessile. The many staminate heads, on both the main axis and the branches, are white and flower successively upward. Stamens 1 cm long, with anthers tapering, bluntly club-shaped, and over 1 mm long. The pistillate flower is sessile, with six spatulate tepals more than half as long as the ovary, and a usually bicarpellate ovary with two divergent stigmas 2 mm or more long.

Fruiting heads 2 – 3.5 cm in diameter, light yellowish brown. Fruit a sessile, relatively broad, top-shaped, usually two-seeded achene, up to 1 cm long, with a rounded or subtruncate top narrowing abruptly to a stout beak 3 – 4 mm long; the stigmas, when persisting, adding a further 2 – 3.5 mm to the length. 2n = 30. Figure 69.

Key to Varieties

1a. Top of achene broadly rounded to truncate, with two stigmas. .... var. *eurycarpum*

1b. Top of achene more narrowly rounded or even tapering, commonly with only one stigma. .... var. *greenei*

Note on the Nomenclature

The combination *Sparganium erectum* may apply in part to this species. Linnaeus (1753, p. 971), with a broader concept of a species than is held by modern botanists, included both branched and unbranched plants within his *S. erectum*, indicating that he considered the branched kind to be the typical one. Modern European botanists, such as Clapham, Tutin and Warburg (1962), Cook (1980) and Stace (1997) accept this statement in *Species Plantarum* as definitive and apply the name to the European relatives of our *S. eurycarpum*. The fact that the only specimen identified as *S. erectum* in the Linnaean Herbarium (Savage 1945) is an unbranched representative of another species raises questions as to the application of the name *S. erectum*. See also the note under *S. angustifolium* ssp. *emersum*.

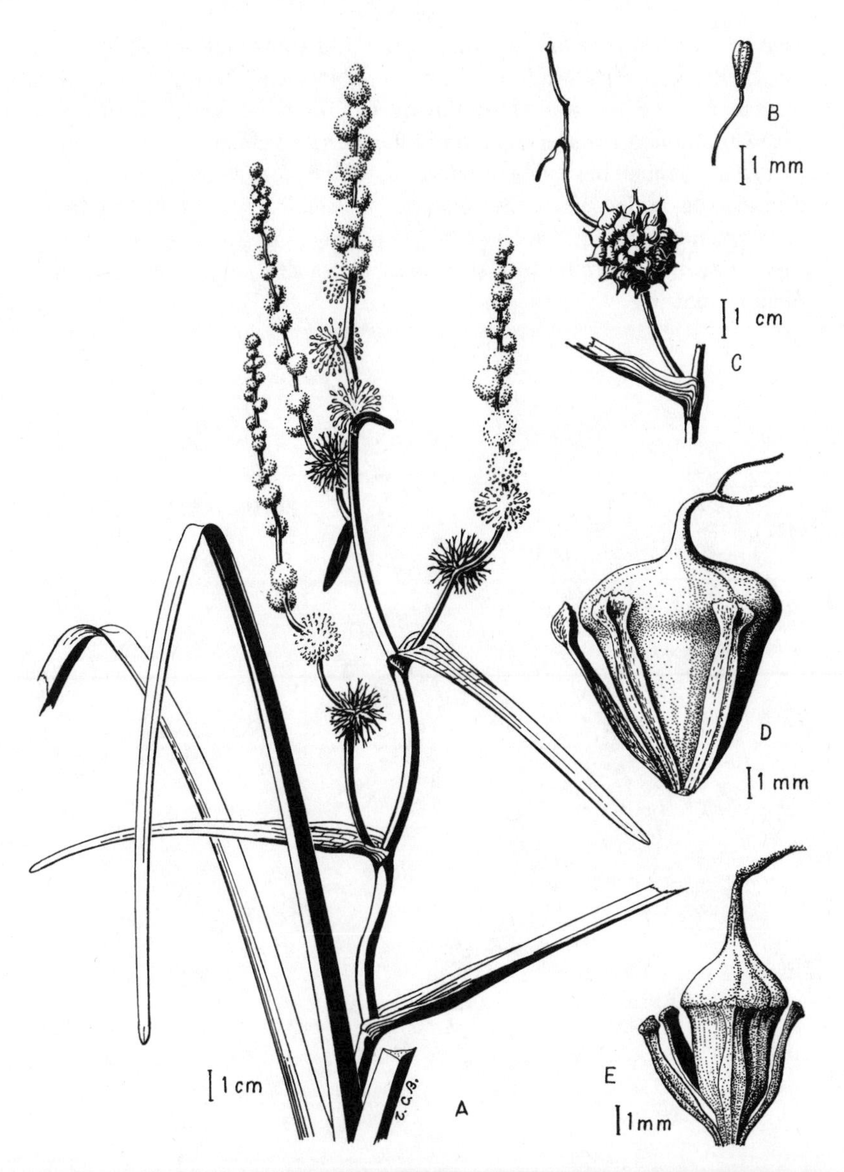

Figure 69. *Sparganium eurycarpum*: A, flowering shoot; B, stamen; C. fruiting head; D, fruit (achene with tepals) of var. *eurycarpum*; E, fruit of var. *greenei*.

Sparganium eurycarpum is widespread and transcontinental in North America, reaching northward into the western Northwest Territories, eastward to Prince Edward Island and southward to Oklahoma and California. In British Columbia it is more common in the interior than at the coast. Variety *eurycarpum* is found in shallow water, 20 – 100 cm deep, around lake margins and in marshes, often where the soils are clays, calcareous or slightly basic. Variety *greenei* (Morong) Graebner ranges mainly along the coast from southwestern British Columbia to California, with a few inland records, as at Williams Lake.

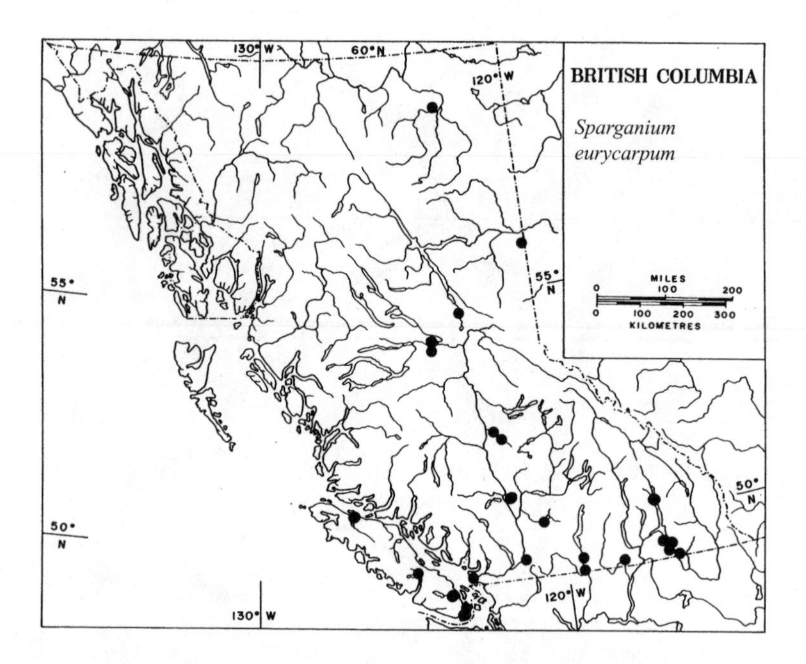

Sparganium fluctuans (Morong) Robinson — Streaming Bur-reed

S. androcladum **(Engelmann) Morong var.** ***fluctuans*** **Morong**

Perennial with slender flexible stems up to two metres long, reclining in the water and most of it usually submerged; the stout, buoyant upper end lying just beneath the surface, and supporting the emergent inflorescence on a stout flexure.

Leaves elongate, ribbonlike, 2 – 11 mm wide, commonly floating, or the upper ones sometimes ascending and emergent, with their bases moderately dilated.

Inflorescence emerging obliquely from the water, branching, the lowest bract shorter to longer than the inflorescence. The inflorescence bearing axillary or sometimes supra-axillary pistillate heads below, and one to four staminate heads above on each branch as well as on the main axis. Stamens 4 – 5 mm long, their anthers elliptic, 0.4 – 0.8 mm long. Pistillate flowers with tepals, often hooded at the tips, originating about the middle of the stipe or below, and barely reaching the middle of the ripe achene. Stigmas ovate to lanceolate, 0.4 – 0.7 mm long.

Fruiting heads 1.3 – 2.3 cm in diameter. Achenes 6 – 10 mm long – including the short stipes and strongly curved, rather flat beaks (2 – 3 mm long) – and often with a median constriction; the lower part of the achene is surrounded by persistent tepals. 2n = 30. Figure 70.

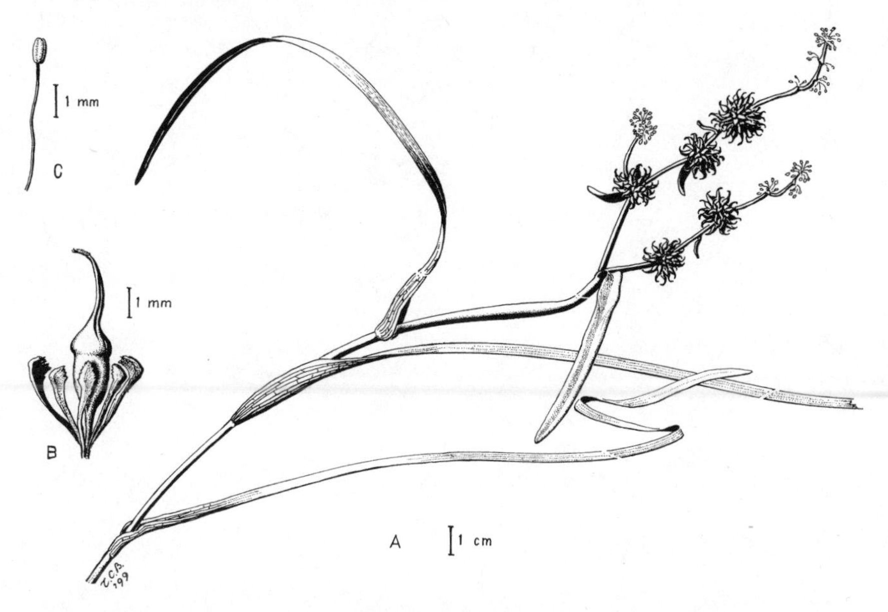

Figure 70. *Sparganium fluctuans*: A, upper, floating part of plant, and fruiting inflorescence; B, achene with tepals; C, stamen.

Found in cold ponds, lakes and slow rivers, often in relatively deep water (1 – 2 metres) for this genus, *Sparganium fluctuans* ranges across boreal North America from British Columbia to Newfoundland and the northern United States. This species is uncommon in central and southwestern British Columbia, including Vancouver Island and the Queen Charlotte Islands.

From their descriptions, this species appears to be closely similar to *S. gramineum* Georgi, of northern Europe (Cook 1980, p. 274), of which our plant may be found to be a North American subspecies after critical comparison of the plants themselves.

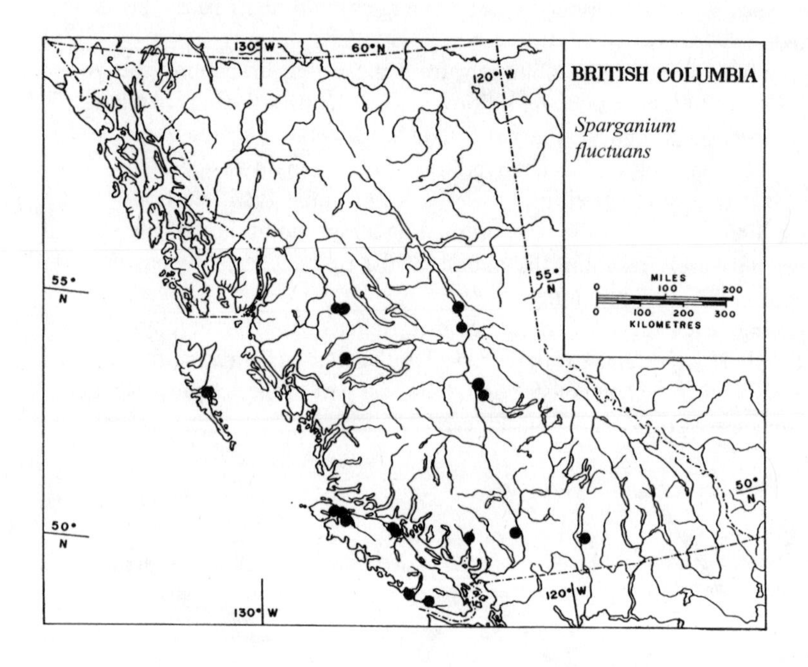

Sparganium glomeratum Laestadius

Stem rather stout, 10 – 60 cm long, upstanding or floating. Leaves stouter than in *S. hyperboreum*, 3 – 20 mm wide, floating or the upper ones emergent with the inflorescence.

Inflorescence with the lowest pistillate head distinct and supra-axillary, and the others aggregated at the top of the stem. Staminate head usually only one, and so close to the upper pistillate head that it appears to be sunk into it at the fruiting stage.

Fruiting heads 1 – 2 cm in diameter. Achenes 5 – 7 mm long overall, tapering into a short stipe and a conical beak, 1 – 1.7 mm long. Scale-like tepals linear to oblanceolate, reaching about the middle of the achene.

A plant of water margins and shallow water, this is primarily an Old World species. Rare in North America, *S. glomeratum* has been recorded at scattered locations from Yukon to Alberta to Quebec and Minnesota. Two collections from British Columbia were noted in the first edition of this book (see the map on page 159 of that edition). The specimen from Kathlyn Lake, Smithers, on which I partly based the illustration of the species (figure 65 in the first edition), has been re-identified by V.L. Harms as *S. angustifolium*, a change that I support. As for the immature specimen from Chaatl Island, Queen Charlotte Islands, on a recent visit (1997) to the University of British Columbia, where the specimen was stored, I could not find it, so I presume that it too has been re-identified.

At present, I know of no confirmed record of *S. glomeratum* from British Columbia. I have retained the description in this edition, in case this species is again found in the province, I omit the illustration, because in the absence of confirmed examples of *S. glomeratum*, I have no specimen on which to base a drawing.

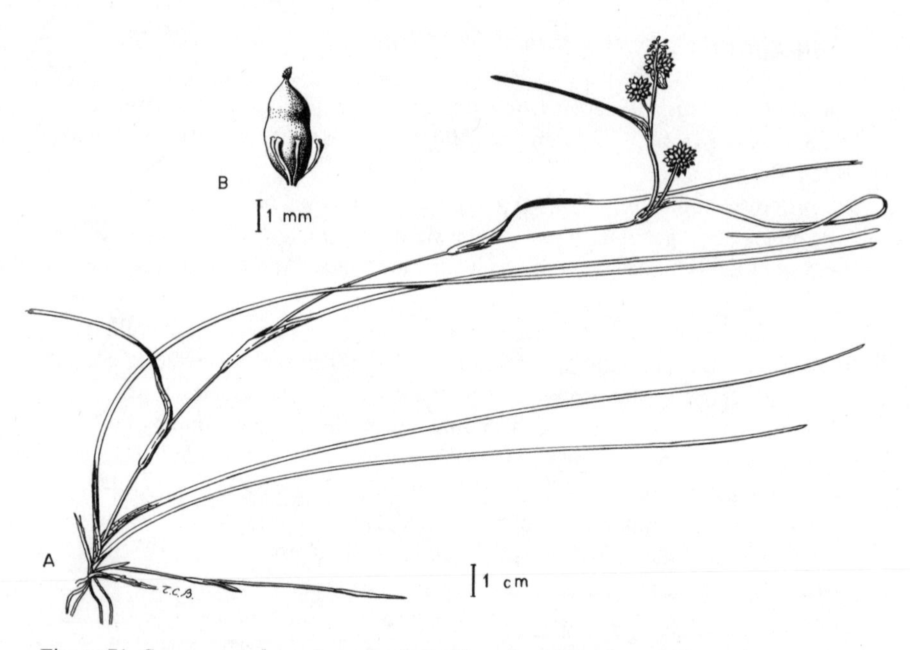

Figure 71. *Sparganium hyperboreum*: A, fruiting plant; B, achene with tepals.

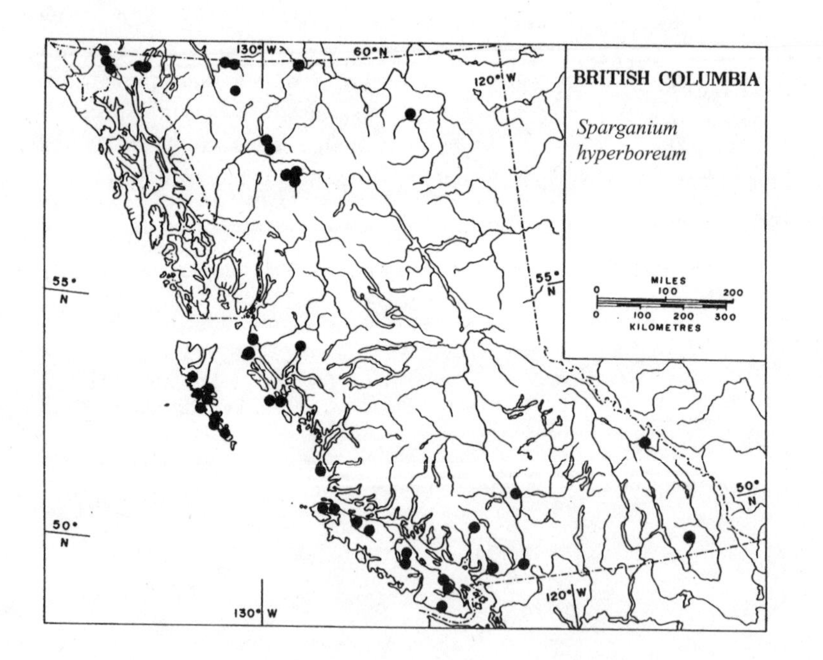

Sparganium hyperboreum Laestadius *ex* Beurling — Northern Bur-reed

***S. natans* L. var. *submuticum* Hartman**

Typically floating, or submerged with only the inflorescence emergent. Stem very slender and weak, and rhizomes similar. Leaves flaccid, 10 – 50 cm long by 1 – 5 mm wide, relatively thick and opaque, with funnel-like or tubular sheathing bases, at least when young.

Inflorescence unbranched, with few heads. At least some pistillate heads supra-axillary, the lowest usually peduncled, the uppermost close to the usually single staminate head. Stamens 5 – 8 mm long, sessile or on a short beak up to 0.5 mm long. Tepals spatulate, a third to half as long as the ripe ovary.

Fruiting heads 5 – 13 mm in diameter. Achenes about 4 mm long, beakless or with a short conical beak up to 0.5 mm long. 2n = 30. Figure 71.

This circumpolar species reaches the Arctic Coast and Greenland in North America. It is widespread in British Columbia, commonly at high altitudes in the interior, but descending to near sea level on the coast. It is found in shallow water, to 0.5 metre deep, commonly with a muck bottom. Harms (1973) has recorded putative hybrids with *S. natans* (as *S. minimum*) from widely scattered locations between Alaska and Newfoundland, but none from British Columbia.

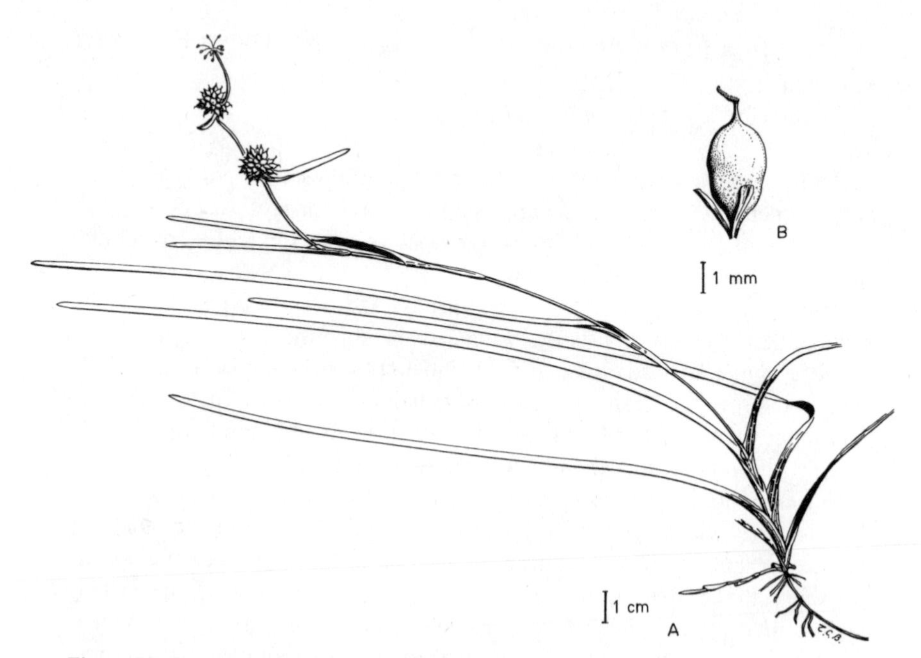

Figure 72. *Sparganium natans*: A, fruiting plant; B, fruit (achene with tepals).

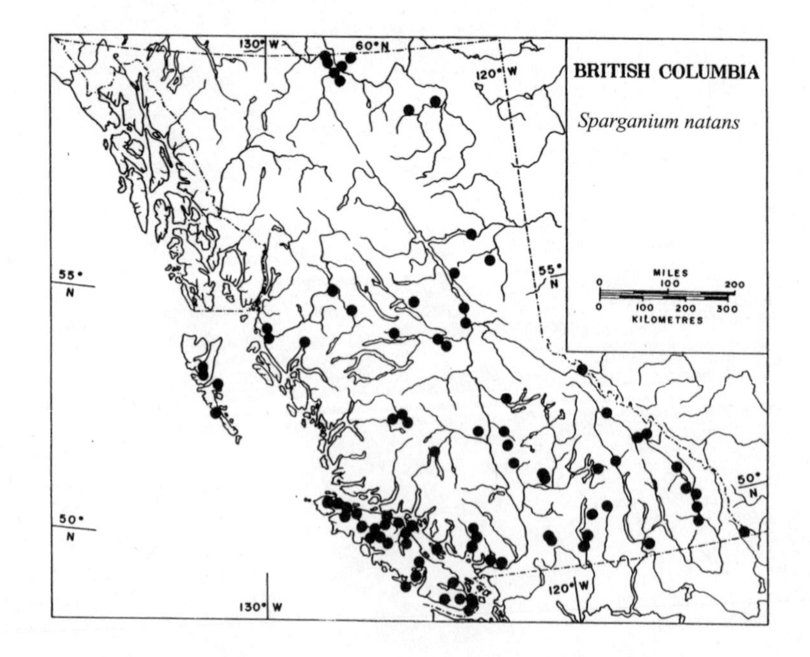

Sparganium natans L. **Floating Bur-reed**

***S. natans* var. *minimum* Hartman**
***S. minimum* (Hartman) Wallroth (1840)**
***S. minimum* (Hartman) Fries (1849)**

Similar in habit to *S. hyperboreum*, but with leaves 2 – 8 mm wide, flatter, relatively thinner, and often translucent, from sheathing bases with margins coherent when young.

Inflorescence unbranched, with all pistillate heads axillary and normally sessile, though occasionally, peduncled, supra-axillary heads are found. Staminate heads one (or rarely two), well separated from the uppermost pistillate heads. Stamens as in *S. hyperboreum*, the anthers 0.8 mm or less long, and nearly as wide as its length. Pistillate flower with spatulate tepals about half as long as the fruit. Stigma elongate, up to 0.8 mm long, on a beak 0.5 – 1.5 mm long.

Fruiting heads 5 – 15 mm in diameter. Achenes 3 – 4 mm long, with a distinct beak up to 1.5 mm long. 2n = 30. Figure 72.

The habitat of *S. natans* is similar to that of *S. hyperboreum*, being found in water 10 – 60 cm deep. *S. natans* is a circumboreal species (Hulten 1941, vol. 1), with a rather more southern distribution than *S. hyperboreum*. In North America it extends from Alaska to Newfoundland and Greenland, and southward to New York, Colorado and northern California (Harms 1973). This species is widespread across British Columbia. A large, apparently empty area in the north and northwest of the province may be real or it may reflect the scarcity of collectors in that region.

Harms (1973) found indications of hybridization between *S. natans* and *S. hyperboreum* in areas where they both grow, in Alaska and at scattered points across Canada, but none in this province.

Note on the Nomenclature

From the entry on *Sparganium natans* L. (Linnaeus 1753, page 971), this combination could be applied either to *S. hyperboreum* or to *S. minimum*. Reference to the Linnaean Herbarium (Savage 1945) as photographed, reveals only one specimen identified as *S. natans*. This specimen is clearly *S. minimum*, as we understand it, with all pistillate heads axillary, and the staminate head well separated from the uppermost pistillate head. From this, it is concluded that the Linnaean combination *Sparganium natans* is the one that should be applied to the species commonly referred to as *S. minimum*.

GLOSSARY

Several plant structures described below are illustrated in figure 73, following the glossary on page 238.

Achene – Dry (non-fleshy) fruit that does not open to release its single seed. Figure 13B.

Acuminate (leaf apex) – Narrowly tapering to acute tip.

Adnate – Joined together, as applied to unlike organs (e.g., the tepal and stamen in *Potamogeton perfoliatus*).

Alternate (leaves) – Leaves arising singly at nodes, the usual arrangement in monocotyledons.

Anther – Pollen-containing body, often compound; the essential part of a stamen. Figure 73.

Anther-sac – Pollen-containing chamber in an anther; also called a pollen-sac.

Apical – Placed on the end of a stem, branch, leaf or fruit.

Appressed – Lying close to or pressed flat against the surface.

Asexual (reproduction) – Reproduction through non-sexual, or vegetative means, such a with rhizomes, tubers or winter buds.

Attenuate – Tapering gradually to a drawn-out tip.

Axil – Angle or gap above the point of attachment of a leaf or leaf stalk, and between that and the stem. Figure 73.

Axillary scales – Minute scales in the axil of a leaf or leaf stalk, and attached to their bases; called *squamulae intravaginales* in technical manuals. See figure 5.

Bipinnate – Doubly pinnate.

Bisexual (flower) – Having both functional stamens and functional ovaries or carpels.

Bog – A wetland community dominated by *Sphagnum* mosses. The accessible soil is spongy, moss-derived peat and the water is usually acidic and low in nutrients.

Bract – A leaf, commonly reduced or otherwise modified, in the axil of which a flower or flower cluster arises. Figure 73.

Caducous – Shedding early in the growing season.

Cambium – A kind of meristem, forming a layer of growing tissue that generates conductive tissues in stems of dicotyledons.

Carpel – Greatly modified leaf, enclosing one or more ovules or seeds; the fundamental unit of the pistil, ovary or fruit. Figure 73.

Chromosome – A linear body (one of several in each cell), whose essential component is a linear molecule of deoxyribonucleic acid (DNA), which carries the code of genes that determine the inheritable characters of an individual.

Circumboreal – Distributed all around the northern hemisphere in cool temperate latitudes.

Circumpolar – Distributed around the northern hemisphere in Arctic latitudes.

Clone – One of a population of genetically identical individuals generated through an asexual, vegetative means of reproduction, such as by rhizomes, tubers, offsets or detached buds.

Connate – Joined together, as applied to similar organs (e.g., carpels in the flowers of *Triglochin*).

Cordate – Heart-shaped, as applied to leaf blade or leaf base (as in *Calla*, figure 56).

Cortex – Relatively unspecialized ground tissue in stems or rhizomes, in which vascular strands are commonly embedded, and which may be used for storage of food reserved. Its cells are often separated by intercellular air spaces (lacunae). (See *Potamogeton praelongus*, figure 40C.)

Connective – The terminal part of the filament of a stamen, connecting the anther-sacs.

Cosmopolitan – Found in all countries around the world.

Cotyledon – Seed leaf; the first one or two leaves formed in an embryo, functioning for nutrient storage in seeds without endosperm, or for nutrient transfer in seeds with endosperm. At germination, they may or may not emerge from the seed to become photosynthetic.

Crenate (leaf margin) – With rounded teeth.

Cuneate (leaf base) – Wedge-shaped, tapering with straight edges.

Cuticle – Waxy (or oily) waterproofing film coating the surface of the epidermis.

Cymose – Having the developmental characteristics of a cyme, an inflorescence in which the terminal flower is older than the nearby lateral flowers and opens before them (e.g., cymose panicle of *Alisma plantago-aquatica*).

Dentate (leaf margin) – With teeth that are equilateral or not regularly oblique.

Dioicous – Having functional sexual structures (stamens and ovaries) on separate plants.

Diploid (chromosome number) – The doubled complement of chromosomes as found in the cells of all growing plant tissues, composed of one basic, haploid set inherited from each parent through sexual (as opposed to vegetative) reproduction. It can also be applied to a species or population possessing two sets of the basic haploid chromosome complement (n or x) for the group.

Dorsal keel – Keel or ridge on the edge of a fruit facing away from the flower axis. Figure 73.

Drupe – Stone fruit; a fruit in which the seed is contained in a hard shell, which in turn, is embedded in a fleshy outer husk.

Ellipsoid – Three-dimensional oval form (like a chicken's egg), thickest at about the midpoint.

Emarginate (leaf) – Having a small marginal notch (as in the abraded leaf tips of *Phyllospadix*).

Embryo – The dormant embryonic plant within the seed. Figure 73.

Endosperm – Food storage tissue accompanying, but not part of, the embryo in the seed.

Entire (leaf margin) – Smooth edged; lacking teeth.

Epidermis – Outermost cell layer covering a plant (may be sloughed off older stems and roots).

Eutrophic (water) – Having a rich concentration of dissolved nutrient chemicals.

Extrorse (anther) – Attachment placing the anther on the outer side of its supporting filament, away from the axis of the flower, and opening outward.

Falcate – Sickle-shaped (e.g., submerged leaf of *Potamogeton amplifolius*, as seen from the side, figure 24A).

False umbel – An apparent umbel, formed from a basically alternate branching pattern through failure of two or more internodes to elongate. Figure 2A.

False whorl – An apparent whorl formed from a basically alternate branching pattern through failure of two or more internodes to elongate. Figure 13A.

Fen – Open wetland on a consolidated surface dominated by sedges or grasses; trees or shrubs are sparse or absent.

Filament – The part of a stamen supporting the anther. Figure 73.

Filiform (leaf) – Threadlike leaf that may be nearly as thick as its width. Figure 37A.

Follicle – A fruit formed of a carpel that ultimately splits open to release its seeds. Figure 2D.

Fruit – Seed-containing body; the product of the flower or ovary, or sometimes of a whole inflorescence.

Glabrous – Without hairs; smooth.

Gland – Small body, often projecting from a plant's surface, commonly for the secretion of gums or other substances, but its function is not always known. Figure 73.

Globose – Approximately spherical.

Haploid (chromosome complement or number) – The basic, indivisible set of chromosomes, as found in the egg, pollen grain, or sperm.

Head (inflorescence) – Dense globular aggregate of sessile or subsessile flowers. Figure 69A.

Hexaploid – Having six complete sets of the chromosomes characteristic of a genus.

Hybrid – The offspring from an event of cross-fertilization between individuals of different species.

Hypanthium (plural: hypanthia) – cuplike or tubular basal part of a flower, formed of the joined bases of the sepals, petals and stamens. Figure 63C.

Indehiscent (fruit) – Not opening to release seed (e.g., achene).

Inferior ovary – Ovary immersed in, and joined to, the base of a flower, thus appearing to be below, or inferior in position to, the flower.

Inflorescence – A definitely structured cluster of flowers.

Internode – Segment of a stem between two successive nodes. Figure 73.

Introgression – The introduction of a gene to one gene complex from another, as in introgressive hybridization.

Introrse (anther) – Attachment placing the anther of the inner side of its supporting filament, toward the axis of the flower, and opening inward.

Keel – Longitudinal ridge on a leaf (figure 69A) or fruit (figure 14D).

Lacunae (singular: lacuna) – Air chambers or passages among cells of cortical tissues of stems or of mesophyll in leaves. They may replace xylem in many aquatic plants.

Lacunate tissue – Spongy tissue, containing extensive air passages (lacunae), developed in the cortex of a stem or in the mesophyll of a leaf, by enlargement of intercellular spaces, or by dissolution of cells, to facilitate the diffusion of gases throughout an aquatic plant. Figure 40C.

Lanceolate (leaf) – With a narrow, tapering outline, like the blade of a lance, widest at or below the midlength. Figure 34A.

Lignification – Process of formation of woody tissue (xylem), or of conversion of soft tissue into woody tissue, by the deposition of lignin in cell walls.

Lignified – Made woody in texture by deposition of lignin in cell walls.

Ligular sheath – Membranous upgrowth, originating on the ventral side of a leaf or petiole close to its base; and sheathing the stem and leaf axil. Figure 73.

Ligule – Projecting upper part of ligular sheath above the point of separation of the sheath from the leaf or petiole. Figure 73.

Linear (leaf) – Long, narrow and parallel-sided. Figure 46A.

Locule – Ovule-containing chamber in an ovary; later, the seed-containing chamber in the fruit.

Marsh – Open, treeless wetland on unconsolidated ground, dominated by sedges, bulrushes or cat-tails. May be tidal or freshwater.

Meristem – A growing tissue, where cell division and multiplication produce cells that subsequently develop into mature tissues to be added to the stem or leaf. A common feature of monocotyledons is basal meristem, which occurs at the base of an internode or leaf blade, so that the internode or leaf elongates from its base.

Mesophyll – Green internal tissue in a leaf containing chlorophyll; specialized for functioning in photosynthesis.

Midvein – The relatively strong vein running up the middle of a leaf blade. Figure 12A.

Monoicous – Having functioning sexual structures of both kinds (stamens and ovaries) on the same plant, though not necessarily in the same flower.

Muck – Sediment or wet soil with a very high content of decomposing organic residue.

Mucronate (leaf tip) – With a small abruptly projecting tip on an otherwise rounded apex. Figure 46D.

Net-veined (leaf) – With largely non-parallel veins forming a netted pattern; typical of dicotyledons, but unusual in monocotyledons. Figure 57A.

Node – The point on a stem where a leaf or bract is, or has been, attached; normally a point of departure on a branch. Figure 73.

Nodal gland – Any gland at the node of a stem. Figure 73.

Obovate – Oval (two-dimensional) that is widest near the distal end.

Obovoid – Egg-shaped body (three-dimensional) that is thickest near the distal end.

Opposite (leaves) – Arising opposite each other in pairs at each node.

Orbicular – Circular in outline (two-dimensional).

Ovary – The lower, ovule-containing (and, subsequently, seed-containing) part of a pistil.

Ovate – Oval (two dimensional) that is widest near the basal end.

Ovoid – Egg-shaped body (three-dimensional) that is thickest near the basal end.

Ovule – Egg-containing body in the ovary. After pollination it develops into the seed.

Palmate (leaves) – Shaped like an open hand or palm; having three or more parts (lobes or veins) radiating from one point, often the junction with the petiole.

Panicle – Diffusely branching inflorescence. Figure 12A.

Parallel-veined (leaf) – With veins lying closely parallel to each other, and inconspicuous cross-veins that do not obviously form a netlike pattern. This is the usual pattern in monocotyledons.

Peat – Spongy-textured organic soil composed of plant remains, principally of *Sphagnum* moss; commonly acidic and low in nutrients. The typical soil of bogs.

Pedicel – Stem bearing a single flower. Figure 73.

Peduncle – Stem bearing an inflorescence. Figure 73.

Pedunculate (inflorescence) – Borne on a peduncle.

Peltate – Shaped like a shield or umbrella, supported on a centrally attached stalk (stigmas of *Zannichellia palustris*, figure 53D).

Perfect (flower) – Bisexual, having functional pistil and stamens in one flower.

Perennating – Surviving over winter to the next growing season by vegetative (i.e., non-sexual) means.

Perianth – Collectively the outer (or lower) sterile whorls of a flower (i.e., the sepals and petals, or the tepals).

Petal – Member of the inner whorl of sterile appendages (the perianth) of a flower.

Petiole – Leaf stalk. Figure 73.

Photosynthesis – Natural synthesis of organic compounds from inorganic ingredients (water and carbon dioxide), using the energy of light.

Pinnate – Resembling a feather, with similar lateral appendages attached successively along a central axis.

Pistil – The inner ovule-bearing (and, subsequently, seed-bearing) part of a flower, consisting of the ovary, stigma(s) and often a style, and formed of a carpel or a number of coherent carpels.

Pistillate (flower) – In which only the carpels and ovary function; roughly equivalent to “female” in animals; can also refer to a plant possessing only pistillate flowers.

Plumose – With projecting hairs, featherlike. Figure 18C.

Pollen – Dust-sized grains containing microscopic male plants enclosed in strong, weather-resistant cell walls; borne in the anthers of a flower and released for transport to the stigmas of other flowers.

Pollination – Transfer and reception of pollen with the object of fertilization.

Prophyll – In many monocotyledons, the first leaf on an axillary branch, usually reduced in size, often scale-like, situated in the angle between the branch and the main stem from which the branch has arisen (see *Zostera marina*).

Proteinaceous – Having proteins as the principal ingredients.

Puberulent – With very fine, soft, more-or-less erect hairs.

Raceme – Inflorescence with pedicels of equal-length arising successively along a central axis. Figure 19A.

Raphides – Minute clusters of needle-like crystals of calcium oxalate in the cells of Araceae and Lemnaceae.

Reflexed – Bent sharply and abruptly back and downward (sepals of *Egeria densa*, figure 3A).

Rhizome – A creeping, rooting stem, usually buried and horizontal, from which erect leafy and flowering stems arise. Figure 73.

Rosette – Cluster of leaves arising from a root-crown or rhizome.

Sagittate – Shaped like an arrow-head (e.g., leaf of *Sagittaria latifolia*, figure 14A).

Scape – A leafless stem arising from the base of a plant, supporting a flower or an inflorescence. Figure 10A.

Scapose – Bearing a flower or an inflorescence on a scape.

Schizocarp – A fruit that at maturity splits into one-seeded carpels. Figure 19D.

Sclerenchyma – Supportive tissue occurring as strands of elongate cells with thick cellulose walls, providing mechanical strength without conduction of water or nutrients.

Sepal – One of the sterile appendages of a flower, outside or below the petals.

Serrate (leaf margin) – With a row of oblique teeth, like the teeth of a saw.

Serrulate – With fine oblique sawlike teeth. Figure 43C.

Sessile – Attached directly by its base without a stalk.

Spadix – Flower spike that is thick and commonly fleshy, often with the flower bases embedded in it, the whole forming a fruit. Figure 73.

Spathe – Expanded bract or pair of bracts surrounding and displaying the spadix. Figure 73.

Spatulate – Elongate and narrow, and broadened near the apex (tepals of *Sparganium eurycarpum*, figure 69D).

Spike – Inflorescence in which sessile or subsessile flowers are attached to an elongate central axis (figure 32). A compound spike is a spike of spikes.

Stamen – Floral appendage that bears pollen.

Staminal column – Column composed of a number of stamens with their filaments joined. Figure 7C.

Staminate (flower) – In which only the stamens function; roughly equivalent to "male" in animals; can also apply to a plant possessing only staminate flowers.

Staminode – Sterile, non-functional stamen, commonly lacking an anther. Figure 4C.

Stigma – Apical or near-apical surface on a pistil or carpel, specialized for the reception of pollen. Figure 73.

Stipe – Elongated stemlike base of a carpel or ovary. Figure 47C.

Stipitate – Elevated on a stipe.

Stipules – Paired appendages associated with leaf or petiole bases in many dicotyledons.

Stolon – Horizontal stem running over the ground or lake bed, enabling a plant to spread and form a colony. Figure 7A.

Stoloniferous – Bearing and spreading by stolons.

Stomata (singular: stoma) – Pores in the leaf epidermis that permit the diffusion of gases between internal tissues and the outside air.

Style – Elongated extension of an ovary or carpel, bearing the stigmas on its upper parts. Figure 16C.

Stylar beak – Beak on a fruit, formed from the style. Figure 73.

Sub- (prefix) – Not quite, almost, as in subopposite (almost opposite) or subsessile (not quite sessile).

Succession – A natural process in which one plant community replaces another over time.

Superior ovary – Ovary placed above and separate from the bases of the flower and of its appendages.

Supra-axillary (branch) – Appearing to depart from the main stem above a leaf axil, due to a union of the base of the axillary branch with the stem above the leaf axil (as in the pistillate peduncles in *Sparganium angustifolium*, figure 67A).

Suture – The seam or edge of a carpel, where splitting may occur to release the seeds. The ventral suture is facing the axis of the flower and is formed of the appressed margins of the carpel.

Swamp – Wetland community dominated by trees or shrubs; its water is not as acidic as a bog.

Tepal – Sterile perianth appendage that is not differentiated as a sepal or a petal. Figure 73.

Terete – Cylindrical and circular in cross-section.

Tetrad – Group of four coherent pollen grains. Figure 65D.

Three-ranked (leaves) – Alternate and in three longitudinal ranks along a stem, every third leaf standing in the same rank; common in monocotyledons.

Tracheae – Tubular vessels in the xylem, specialized for water conduction, and composed of lines of large cells placed end to end, with the intervening partitions perforated, and the walls at least partly lignified. Characteristic of vascular systems in most terrestrial flowering plants.

Trimerous (flower) – Having its floral appendages in whorls of three.

Triploid (chromosome number) – Having three sets of the basic complement of chromosomes (see Diploid). Triploid plants are usually sterile.

Truncate (leaf apex) – Broadly rounded to almost square, or appearing as if cut off. Figure 49A, B.

Two-ranked (leaves) – Arising in two ranks along a stem, alternately to left and to right. Figure 39A.

Type specimen – A specimen used for the description of a species or variety, or identified as a typical example by the describer or monographer.

Umbel – Inflorescence in which all branches arise from one node at the top of the peduncle.

Undulate (leaf margin) – Wavy or undulating above and below the plane of the leaf. Figure 25A.

Unicarpellate – Having only one carpel.

Unilocular (ovary) – Having only one locule.

Unisexual (flower) – In which only stamens or carpels function, but not both.

Utricle – Thin-walled, often bladderlike fruit containing a single seed. Figure 52D.

Vascular strand – A clearly defined strand of vascular conducting tissues running through the stems, leaves and other organs, for the transportation of water, food, etc. The veins of leaves are vascular strands.

Vascular tissues – Tissues with elongated slender cells, often with thick walls, specialized for the conduction of water, nutrients, etc. In land plants, vascular tissues often help support the structure. They are the essential tissues in stems and leaf veins.

Vegetative shoot – A shoot bearing leaves only, without flowers.

Vegetative reproduction – Plant reproduction without pollination; e.g., by tuber, rhizome or winter bud.

Ventral – Facing the axis of a stem or flower.

Ventral keel – Keel or ridge on the edge of a fruit facing the flower axis. Figure 73.

Vestigial – Reduced to a vestige or trace of an original, larger form.

Villous – With long, soft, curly hairs that are not matted.

Whorl – A ring of three or more leaves or branches around a single node. Figure 4A.

Winged – Having longitudinal flanges or projecting edges (as in the achene of *Potamogeton crispus*, figure 25C).

Winter bud – A bud produced late in the growing season that is shed to lie dormant over winter and sprout in spring to produce a new plant.

Xylem – Tissue specialized for water conduction (and for rigid support in many land plants), with long, slender cells whose walls are commonly stiffened with lignin. Xylem is the fundamental tissue of wood.

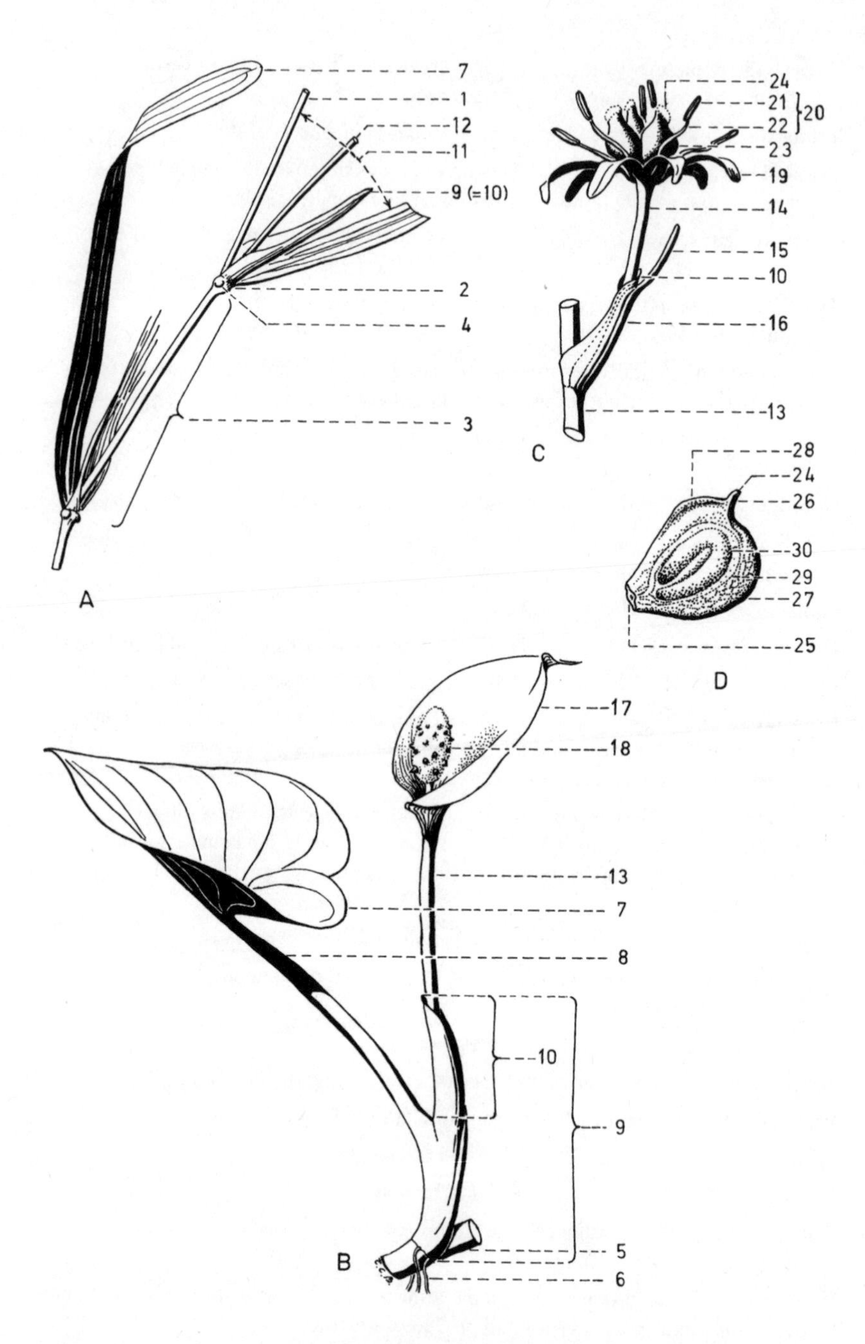

Figure 73. Structures associated with stems, leaves, flowers and fruits (see the key on the facing page): A, stem and leaves of *Potamogeton obtusifolius*; B, leaf and inflorescence of *Calla palustris*; C, flower of *Scheuchzeria palustris*; D, fruit (an achene) of *Sagittaria cuneata*.

Abbreviations and Symbols

cm centimetre(s).
DC. De Candolle, Augustin Pyramus.
ex published by.
L. Linnaeus, Carolus (the elder).
mm millimetre(s).
n the number of chromosomes in the haploid (single) complement, as found in the egg, pollen or sperm of a species, or small population, or an individual; normally equal to or a multiple of x.
2n the number of chromosomes in the diploid (doubled) complement of chromosomes, as found in the cells of vegetative tissues.
p. page number.
sp. species (singular).
spp. species (plural).
ssp. subspecies.
var. variety.
x number of chromosomes in the basic haploid complement for a series of species, or for a genus or larger category.
X crossed with. When between specific names, it designates a hybrid between the two species; when preceding a name, it designates a named hybrid.

Parenthetical Range Limits

Numerical ranges can be written with dashes (e.g., 12 – 24). A range expressed as (9 –) 12 – 24 means that 12 to 24 is usual or common and below 12 to 9 is rare but still possible.

The Generic Initial

Where the name of a genus is used repeatedly in the discussion of a species, it is spelled out the first time and thereafter referred to by its initial, as long as there is no possibility of confusion with another genus. For example, *Alisma gramineum* in the first mention and *A. gramineum* afterward.

Key to Plant Structures Illustrated on the Facing Page

1. Main stem
2. Node
3. Internode
4. Nodal gland
5. Rhizome
6. Root
7. Leaf blade
8. Petiole
9. Ligular sheath
10. Ligule
11. Axil or leaf
12. Axillary branch
13. Peduncle
14. Pedicel
15. Bract
16. Bract sheath
17. Spathe
18. Spadix
19. Tepal
20. Stamen
21. Anther
22. Filament
23. Carpel
24. Stigma
25. Basal scar (point of attachment to flower axis)
26. Stylar beak
27. Dorsal keel
28. Ventral keel
29. Locule
30. Embryo in seed

REFERENCES

Arber, Agnes. 1920. *Water Plants*. Cambridge, U.K.: Cambridge University Press. Reprinted 1963, Weinheim: J. Cramer.

Bayer, Range D. 1979. Intertidal zonation of *Zoster marina* in the Yaquina Estuary, Oregon. *Syesis* 12:147–54.

Boivin, Bernard. 1967. *Enumeration des Plantes du Canada*. Provancheria No. 6. *Naturaliste Canadien* (Université Laval, Quebec) 93:515-519. .

Calder, James A., and Roy L. Taylor. 1968. *Flora of the Queen Charlotte Islands*. Monograph No. 4, Part 1. Ottawa: Research Branch, Agriculture Canada.

Ceska, Adolf, and Oldriska Ceska. 1980. Additions to the Flora of British Columbia. *Canadian Field-Naturalist* 94:69–74.

Charlton, W.A. 1973. Studies in the Alismataceae II: Inflorescences of Alismataceae. *Canadian Journal of Botany* 51:775–89.

Charlton, W.A., and A. Ahmed. 1973. Studies in Alismataceae IV: Developmental morphology of *Ranalisma humile* and comparisons with two members of the Butomaceae, *Hydrocleis nymphoides* and *Butomus umbellatus*. *Canadian Journal of Botany* 51:899–910.

Clapham, A.R., T.G. Tutin and E.F. Warburg. 1962. *Flora of the British Isles*. Second edition. Cambridge, U.K.: Cambridge University Press.

Cody, William J. 1996. *Flora of the Yukon Territory*. Ottawa: NRC Research Press.

———. 1998: Horned Pondweed *Zannichellia palustris* (Zannichelliaceae), new to the vascular plant flora of the continental Northwest Territories, Canada, and deleted from the flora of the Yukon Territory. *Canadian Field-Naturalist* 112(4):711–12.

Cook, C.D.K. 1968. Phenotypic plasticity with particular reference to three amphibious plant species. In *Modern Methods in Plant Taxonomy*, edited by V.E. Heywood. London, U.K.: Academic Press.

———. 1980. Sparganiaceae. In *Flora Europaea*, vol. 5, edited by T.G. Tutin, V.H. Heywood, N.A. Burges, D.M. Moore, D.H. Valentine, S.M. Walters and D.A. Webb. Cambridge, U.K.: Cambridge University Press.

Cronquist, Arthur. 1988. *The Evolution and Classification of Flowering Plants*. Bronx: New York Botanical Garden.

Cronquist, A., A.H. Holmgren, N.H Holmgren, J.L. Reveal and P.K. Holmgren. 1977. *Intermountain Flora: Vascular Plants of the Intermountain West, U.S.A.*, vol. 6. New York: Columbia University Press.

Dacey, J.W.H. 1980. Internal winds in water lilies: an adaptation for life in anaerobic sediments. *Science* 210:1017–19.

Dahlgren, R.M.T., H.T. Clifford and P.F. Yeo. 1985. *The Families of the Monocotyledons*. Berlin: Springer Verlag.

Dale, H.M., and T.J. Gillespie. 1977. The influence of submerged aquatic plants on temperature gradients in shallow water bodies. *Canadian Journal of Botany* 55:2216–25.

Dandy, J.E. 1980a. Alismataceae. In *Flora Europaea*, vol. 5, edited by T.G. Tutin, V.H. Heywood, N.A. Burges, D.M. Moore, D.H. Valentine, S.M. Walters and D.A. Webb. Cambridge, U.K.: Cambridge University Press.

———. 1980b. *Triglochin*. In *Flora Europaea*, vol. 5, edited by T.G. Tutin, V.H. Heywood, N.A. Burges, D.M. Moore, D.H. Valentine, S.M. Walters and D.A. Webb. Cambridge, U.K.: Cambridge University Press.

———. 1980c. *Potamogeton*. In *Flora Europaea*, vol. 5, edited by T.G. Tutin, V.H. Heywood, N.A. Burges, D.M. Moore, D.H. Valentine, S.M. Walters and D.A. Webb. Cambridge, U.K.: Cambridge University Press.

Dandy, J.E., and G. Taylor. 1938–1942. Studies of British *Potamogetons* I – XVIII. *Journal of Botany*, vols 76–80. London: Linnaean Society.

Daubs, Edwin Horace. 1965. *A Monograph of Lemnaceae*. Illinois Biological Monographs 34. Urbana: University of Illinois Press.

Den Hartog, C. 1970. *The Sea-Grasses of the World*. Amsterdam: North Holland.

Dore, William G. 1957. *Wolffia* in Canada. *Canadian Field-Naturalist* 71:10–16.

Douglas, George W., Gerald B. Straley, and Del Meidinger. 1994. *The Vascular Plants of British Columbia*, Part 4: *Monocotyledons*. Special Report no. 4. Victoria: British Columbia Ministry of Forests.

Fassett, Norman C. 1957. *A Manual of Aquatic Plants*. Second edition revised by Eugene C. Ogden. Madison: University of Wisconsin Press.

Fernald, M.L. 1921. Expedition to Nova Scotia. *Rhodora* 23:184–95.

———. 1950. *Gray's Manual of Botany*. Eighth edition. New York: American Book Co.

Gleason, Henry A., and Arthur Cronquist. 1991. *Manual of Vascular Plants of Northeastern United States and Adjacent Canada*. Second edition. Bronx: New York Botanical Garden.

Harms, Vernon L. 1973. Taxonomic studies of North American *Sparganium*, 1: *S. hyperboreum* and *S. minimum*. *Canadian Journal of Botany* 51:1629–41.

Harrison, Paul Garth. 1976. *Zostera japonica* (Aschers. and Graebn.) in British Columbia, Canada. *Syesis* 9: 359–60.

Haynes, Robert R. 1974. A revision of North American *Potamogeton*, subsection *Pusilli* (Potamogetonaceae). *Rhodora* 76: 564–649.

Haynes, Robert R., and C. Barre Hellquist. 1996. New combinations in North American Alismatidae. *Novon* 6: 370–71.

Helliquist, C. B., and G.E. Crow. 1980. *Aquatic Vascular Plants of New England*, part 1: *Zosteraceae, Potamogetonaceae, Zannichelliaceae, Najadaceae*. Station Bulletin 515. Durham: New Hampshire Agricultural Experiment Station (University of New Hampshire).

Hitchcock, C. Leo. 1969. All aquatic families of monocotyledons. In *Vascular Plants of the Pacific Northwest*, Part 1, by C. Leo Hitchcock, Arthur Cronquist, Marion Ownbey and J.W. Thompson. Seattle: University of Washington Press.

Hitchcock, C. Leo, and Arthur Cronquist. 1973. *Flora of the Pacific Northwest*. Seattle: University of Washington Press.

Holub, Josef. 1997. *Stuckenia* Borner 1912 – the correct name for *Coleogeton* (Potamogetonaceae). *Preslia* 69: 361–66.

Howell, J.T. 1947. *Triglochin concinna* var. *debilis* (M.E. Jones) J.T. Howell. *Leaflets of Western Botany* 5(1):18.

Hulten, Eric. 1941. *Flora of Alaska and Yukon*, vol. 1. Lund, Sweden: Håkon Ohlssons Boktryckeri.

———. 1968. *Flora of Alaska and Neighboring Territories*. Stanford, California: Stanford University Press.

Jones, M.E. 1895. *Triglochin maritimum* var. *debile* (description). *Proceedings of the California Academy of Science* 2(5):722.

Kingsbury, John M. 1964. *Poisonous Plants of the United States and Canada*. Englewood Cliffs, New Jersey: Prentice-Hall.

Landolt, Elias. 1975. Morphological differentiation and geographical distribution of the *Lemna gibba – Lemna minor* group. *Aquatic Botany* 1(4):345–63.

Les, Donald H. 1983. Taxonomic implications of aneuploidy and polyploidy in *Potamogeton* (Potamogetonaceae). *Rhodora* 85: 301–23.

Les, Donald H., M.A.C. Eland and T.C. Philbrick. 1995. Taxonomic realignments in Potamogetonaceae: evidence from molecular data. *American Journal of Botany* 82 (6) Supplement – Abstracts: 144.

Les, Donald H., and Robert R. Haynes. 1996. *Coleogeton* (Potamogetonaceae), a new genus of pondweeds. *Novon* 6:389-91.

Linnaeus, Carl. 1753. *Species Plantarum*. Facsimile reprint, 1957, London, U.K.: Ray Society.

Lomer, Frank, and George W. Douglas. 1999. Additions to the vascular plant flora of the Queen Charlotte Islands, British Columbia. *Canadian Field-Naturalist* 113(2):235–40.

Löve, Askell, and Doris Löve. 1958. Biosystematics of *Triglochin maritimum* agg. *Le Naturaliste Canadien* 35:156–65.

Packer, John G., and Gordon S. Ringius. 1984. The distribution and status of *Acorus* (Araceae) in Canada. *Canadian Journal of Botany* 62:2248–52.

Phillips, Ronald C. 1979. Ecological notes on *Phyllospadix* (Potamogetonaceae) in the northeast Pacific. *Aquatic Botany* 6:159–70.

Pojar, J. 1973. Levels of polyploidy in four vegetation types of southwestern British Columbia. *Canadian Journal of Botany* 51:621–28.

Porsild, A.E., and W.J. Cody. 1980. *Vascular Plants of Continental Northwest Territories, Canada*. Ottawa: National Museum of Natural Sciences.

Posluszny, U., and R. Sattler. 1976a. Floral development of *Zannichellia palustris*. *Canadian Journal of Botany* 54:651–62.

———. 1976b. Floral Development of *Najas flexilis*. *Canadian Journal of Botany* 54:1140–51.

Reveal, James L. 1970. *Sparganium simplex* Huds., a superfluous name. *Taxon* 19:796–97.

Rowe, J.S. 1972. *Forest Regions of Canada*. Publication no. 1300. Ottawa: Environment Canada, Forestry Service.

Savage, Spencer. 1945. *A Catalogue of the Linnaean Herbarium*. London, U.K.: Linnaean Society of London. (Microfilm.)

Scoggan, H.J. 1978. *The Flora of Canada*, part 2. Publications in Botany no. 7(2). Ottawa: National Museum of Natural Sciences.

Sculthorpe, C.D. 1967. *The Biology of Aquatic Vascular Plants*. London, U.K.: Edward Arnold.

Singh, V. 1965. Morphological and anatomical studies in Helobiae, 2: Vascular anatomy of the flower of Potamogetonaceae. *Botanical Gazette* 126(2):137–44.

Singh, V., and R. Sattler. 1972. Floral development of *Alisma triviale*. *Canadian Journal of Botany* 50:619–27.

———. 1973. Nonspiral androecium and gynoecium of *Sagittaria latifolia*. *Canadian Journal of Botany* 51:1093–95.

———. 1974. Floral development of *Butomus umbellatus*. *Canadian Journal of Botany* 52:223–30.

Spence, D.H.N., and H.M. Dale. 1978. Variations in the shallow water form of *Potamogeton richardsonii* induced by some environmental factors. *Freshwater Biology* 8:251–68.

Stace, Clive. 1997. New flora of the British Isles. Second edition. Cambridge, U.K.: Cambridge University Press.

Stuckey, Ronald L., John R. Wehrmeister and Robert J. Bartolette. 1978. Submersed aquatic vascular plants in ice-covered ponds of central Ohio. *Rhodora* 80:575–80.

Svedelius, Nils. 1932. On the different types of pollination in *Vallisneria spiralis* L. and *Vallisneria americana* Michx. *Svensk Botanisk Tidskrift* 26, H:1–12.

Takhtajan, Amen. 1969. Flowering Plants: Origin and Dispersal. Translated by C. Jeffrey. Edinburgh: Oliver & Boyd.

Taylor, Roy L., and Gerald A. Mulligan. 1968. *Flora of the Queen Charlotte Islands*. Monograph no. 4, part 2. Ottawa: Research Branch, Agriculture Canada.

Taylor, Roy L., and Bruce MacBryde. 1977. Vascular Plants of British Columbia. Vancouver: University of British Columbia Press.

Thorne, Robert F. 1976. A phylogenetic classification of the Angiospermae. *Evolutionary Botany* 9:35–106.

———. 1981. Phytochemistry and Angiosperm phylogeny: a summary statement. In *Phytochemistry and Angiosperm Phylogeny*, edited by David A. Young and David S. Seigler. New York: Praeger Publishers.

Tomlinson, P.B. 1982. Helobiae (Alismatidae). In *Anatomy of the Monocotyledons*, part 7, edited by C.R. Metcalfe. Oxford: Clarendon Press.

Zoltai, S.C. 1976. Wetland classification. *Proceedings of the First Meeting on Ecological (Biophysical) Land Classification* (Petawawa, Ontario): 61–71.

INDEX